Environmental Science

Sri G.V.G Visalakshi College for Women

Autonomous and Affiliated to Bharathiar University

Accredited at "A+" Grade by NAAC

An ISO 9001 - 2015 Certified Institution

Udumalpet - 642128

Published by

BONFRING®
Intellectual Integrity

ISBN 978-93-92537-30-1

Bonfring
309, 5th Street Extension, Gandhipuram,
Coimbatore - 641 012,
Tamil Nadu, India.
E-mail: info@bonfring.org
Website: www.bonfring.org

Foreword

The book entitled "Environmental Science" is written with the view to enlighten the students on an interdisciplinary academic field. It is assumed that all the chapters in this book highlight the prevention of the environmental disruption. It is very essential for the student community to comprehend the greatest challenges that the world face and this book very well aims to promote an understanding of significant environmental issues. The need for the book appears more obvious and is clear that it can develop environment consciousness and conservation of Biodiversity. It creates awareness on environmental pollution, Protection of endangered animals and Environmental audit system etc. The world today makes giant leaps towards progress and let this book serve as a platform to protect environment and support progression.

Mrs.S. Rajalakshmi

Chairman

Nature Science Foundation

Coimbatore.

Preface

The five basic elements of life the "Panch Mahabutas, Akasha, Vayu, Agini, Jalam and Prithvi" are derived from Prakrit, the primal energy. All living and non-living things on this earth are forms of the Supreme Truth, made up of these five elements in varying degrees. These five elements combine with each other and change in these leads to an imbalance in human causing diseases. Man's survival and ability to fight against diseases would depend on how easily he adapts to the environment. Similarly it is extremely important to have an understanding of environmental issues.

"Ecosystems and Humans" serves as an introduction to the broad field of environmental science. It defines environmental science, explains the principles of the ecosystem approach, gives an overview of environmental stressors caused by human activities, and describes various world views.

Environmental literacy can be defined as: "The degree to which people have an objective and well-informed understanding of environmental issues." Today, it is extremely important to have understanding of environmental issues. This is because the human economy is engaged in a wide range of activities that cause enormous damage to the ecosystems. All around us, this is witnessed by pollution, global warming, collapsing fisheries, deforestation, the degradation of agricultural soil, extinctions and endangered plant and animal species, and other damages.

If our society takes constructive actions now, it will not be too late to repair many of these important environmental problems, which threaten the welfare of the people and most other species. Within limits, humans are prescient creatures, and our society is capable of implementing a sustainable economy that can support, our livelihoods as well as healthy ecosystems. It is clear, however, that any sustainable economy will involve ways of doing business that are different from those that have recently been dominant. It will also require fundamental changes in the lifestyles of people. Ultimately, such socio-economic transformations must involve much less use of energy, materials and other resources, in comparison with what many of us take for granted today. A more respectful attitude towards the natural world is also badly needed.

Achieving such a transformation will depend on citizens having a sound understanding of environmental issues. Imposition of restrictions on access to resources will initially be uncomfortable for many people. Nevertheless, in future people will be more willing to soften their lifestyle, if they understand the reasons for those changes in the context of the livelihoods of future generations and ecological sustainability more generally. With such an understanding, most people will support economic and social changes that conserve the quality of their own and future environments. A broad-based environmental literacy will be a key requirement if a country such as India is to achieve the difficult transition into an ecologically sustainable economy. within the context, this book was developed to help our college students to have an objective and well-informed understanding of important environmental issues.

The Editorial board is thankful to the Management and the Principal of our college for their encouragement to publish the new edition of "Environmental Science" Book for I UG students. This book unveils the topics about environmental issues and Environmental Audits. We thank Mrs.S. Rajalakshmi, Chairman, Nature Science Foundation, Coimbatore for her valuable review and suggestions about the book.

Editorial Board

Editorial Board

TABLE OF CONTENTS

CHAPTER I

NATURAL RESOURCES

Introduction

Environmental science is the study of the interactions between the physical, chemical and biological components of the natural world, including their effects on organism and natural resources as well as monuments. Environment is everything that affects an organism during its lifetime.

The word 'Environment' has originated from French word "environ", means surroundings. Environment comprises the surroundings in which man lives, works and plays. Environment is the sum total of abiotic and biotic conditions influencing the response of a particular organism.

The study about environment creates an awareness about the importance of protection and conservation of our mother earth and about the problems arise due to the release of pollution into the environment. The increase in human and animal population, industries and other issues make the survival cumbersome. A great number of environmental issues have grown in size and make the system more complex day by day, threatening the survival of mankind on earth.

In order to preserve our environment, the human beings must be conscious of their activities. In the name of progress and scientific advancement, the environment cannot be sacrificed. As long as there is a balance maintained between the constructive and destructive activities, especially of human beings, a healthy environment can thrive. It is therefore necessary to learn about the environment, our resources, how we depend on them for our survival, and about the various factors which a healthy ecosystem hinge on.

1.1. Environmental Science

Environmental science is a multi-disciplinary science which comprises of Chemistry, Physics, Biological and Agricultural sciences and public health. It is the science of complex interactions among the terrestrial, atmospheric, aquatic and anthropological environment. It is also the study of air, water, soil, living organisms and human beings, and the impact of our activities on the environment.

Scope of Environmental Studies

The scope of the discipline of environmental studies is as follows:

a. It provides knowledge about ecological systems, cause and effect relationship.

b. It provides necessary information about biodiversity, its richness and the potential dangers to the species of plants, animals and microorganisms in the environment.

c. The study enables one to understand the causes and consequences of natural and man induced disasters such as flood, earthquake, landslide, cyclones etc., and pollutions and measures to minimize the effects.

d. The study creates awareness among the people to know about various renewable and non-renewable resources of the region.

e. The study enables environmentally literate citizens (by knowing the environmental acts, rights, rules, legislations, etc.) to make appropriate judgments and decisions for the protection of the environment.

f. It teaches the citizens the need for sustainable utilization of resources without deteriorating their quality.

Importance of Environmental Studies

Study of the Environment becomes significant due to the following reasons:

a. Environmental study helps to understand the various aspects of environment.

b. It emphasizes the importance of conservation of biodiversity.

c. It helps, one to understand the impact of environment on man and his life and the impact of human life on the environment.

d. Most of the natural resources on the earth are limited and are exhaustible. So, there is a need for proper utilization and conservation of natural resources. Environmental study is helpful in addressing this problem and for proper utilization and conservation of limited natural resources.

e. Environmental studies are a key instrument for bringing about the changes in values, behavior and lifestyles required to achieve sustainability and stability within and among nations.

1.2. Types of Resources

Resources are defined as materials which are required for the survival, comfort and prosperity of human beings on the earth.

These sources are classified into two categories–renewable (non-conventional) resources and non-renewable (conventional) resources.

Renewable Resources

Renewable resources are inexhaustible resources, which have the ability to reappear by recycling, reproduction or replacement. They are also called biotic resources. These renewable resources include sunlight, water and living organisms. However, if these resources namely water and living organisms are used faster than the rate at which they can be replenished, they will soon run out.

Non-renewable Resources

Non-renewable resources are those that take millions of years to form naturally and cannot be replaced once they are used. They are called exhaustible resources or abiotic resources. They include minerals and fossil fuels (coal, petroleum, natural gas etc.) that are present in fixed amounts in the environment.

1.2.1. Forest Resources

One of the most valuable natural resources of the world is forest. The term "forest" is derived from the Latin word "foris" meaning outside. In the words of Allen and Shorpe, "Forest is a community of trees and associated organisms covering a considerable area, utilizing air, water and minerals to attain maturity and to reproduce itself and capable of furnishing mankind with indispensable products and services". Originally, forests covered about half of the land area of the world. But today, they cover only one-fourth of the total land area of the earth. The reason for the decrease in the area of land under forest is the destruction of forest for the purpose of habitation and agriculture.

Benefits from Forests

a. Forests are the important habitat for wildlife, where herbivores find their food and shelter and carnivores their prey.

b. Forests play an important role in maintaining the atmospheric carbon dioxide and oxygen balance by consuming the CO_2 present in the air and by producing the O_2 that is very essential for all the living beings.

c. Forests protect the soil from the direct action of rain and wind. As plants hold the soil particles firmly, the erosion is prevented.

d. Plant litter and humus make the soil spongy and the humus holds rainwater and prevent run off. Slowly water percolates into the soil providing perennial supply to springs and rivulets.

e. Forests have a moderating effect on temperature and climatic changes. It increases the rainfall frequencies and humidity of the atmosphere.

f. Plants of the forest are the food producing organisms and are termed as the primary producers of the 'food chain'. Forest serves as an energy reservoir by trapping energy from sunlight and storing it in the form of a biochemical product. The forests provide the human food needs to some extent.

g. Forests are the main source of wood and bamboos that are used as raw material in many industries.

h. Many products such as tannins, gums, drugs, spices, insecticides, waxes, honey, ivory and hides are provided by flora and fauna of forests. Forests also serve as main source of food products such as fruits, leaves, roots and tubers of plants, and meat of forest animals.

Environmental Issues

Deforestation: Deforestation refers to the loss of forest cover or clearing of forests on the surface of the earth. This process involves the felling of trees, burning and damaging of forests. Deforestation occurs due to the following causes.

a. **Shifting Cultivation:** Shifting cultivation in an agricultural system in which plots of land are cultivated temporarily then abandoned and allowed to revert to their natural vegetation while the cultivator moves on to another plot. The length of time that a field is cultivated is usually shorter than the period over which the land is allowed to regenerate by lying fallow.

b. **Hydro-electric Projects:** Dams, reservoirs and hydro-electric projects submerge large forest tracts; they uproot thousands of forest dwellers from their area of residence. A lot of land is cleared for providing residence for the workers; for which wood and other forest products are used up.

c. **Hill and Forest Roads:** Construction of roads and railways in the hilly forest areas brings about a lot of deforestation, which cause landslides and soil erosion.

d. **Forest Fire:** A forest fire which is natural or manmade, continue for several days and destroy a big portion of the forest.

Environmental Effects

The following are the environmental effects of deforestation.

a. **Atmospheric Pollution:** Deforestation is often cited as one of the major causes of the enhanced greenhouse effect. Trees and other plants remove carbon (in the form of carbon dioxide) from the atmosphere during the process of photosynthesis. Both the decay and burning of wood release much of the stored carbon back to the atmosphere. This not only causes atmospheric pollution but also alter the climate of that region.

b. **Water Cycle:** The water cycle is also affected by deforestation. Trees extract ground water through their roots and release it into the atmosphere. When part of a forest is removed, the region cannot hold as much water which may result in a much drier climate.

c. **Biodiversity:** Some forests are rich in biological diversity. Deforestation can cause the destruction of the habitats that support this biological diversity.

d. **Soil Erosion:** India loses some 6000 million tons of top soil every year due to water erosion caused by absence of tree cover in the hills.

e. **Landslides:** Tree roots bind soil together and if the soil is sufficiently shallow, they keep the soil in place by binding with underlying bedrock. Tree removal on steep slopes with shallow soil thus increase the risk of landslides, which can threaten people living nearby.

1.2.2. Water Resources

Water has a unique place among all the renewable resources and it is truly a unique gift to mankind from nature. It is essential for sustaining all forms of life, food production, and economic development for general well-being. Water is also one of the most manageable of the natural resources as it can be transported and stored. Surface water and ground-water resources play a major role in agriculture, hydropower generation, livestock production, industrial activities, forestry, fisheries, navigation, recreational activities, etc.

Surface water is naturally replenished by precipitation and naturally lost through discharge to the oceans, evaporation and sub-surface seepage. India is gifted with a river system comprising more than 20 major rivers with several tributaries. Many of these rivers are perennial and some of these are seasonal. The rivers like Ganges, Brahmaputra and Indus originate from the Himalayas and carry water throughout the year. The snow and ice melt of the Himalayas and the base flow contribute to the flows during the lean season.

About 2.7 percent of the total water available on the earth is fresh water of which about 75.2 percent lies frozen in Polar Regions and another 22.6 percent is present as ground water, the rest is available in lakes, rivers, atmosphere, moisture, soil and vegetation.

Exploitation of Water

World population is rising at an alarming rate, which leads to greater requirements of water for human consumption. There is a great pressure on the water resources of many countries, particularly those developing countries located in arid or semi-arid regions. Acute water shortage exists in many areas of the world and is likely to become more severe in future years. In some areas, further economic growth will not be possible until adequate additional water supplies become available. In other areas, failure to increase the water supply may result in standard of living being reduced below present levels. Competition for water resources will heighten arousing conflicts.

Water Crisis

The reasons for water crisis are:

1. Rapidly increasing population
2. Modern industrialization
3. Agricultural growth

Water is consumed by these factors without any control. Due to deforestation the amount of rainfall is decreased and water is reduced in the rivers, lakes and ponds. Deforestation and urbanization are the main factors for the floods. The surface areas in the cities are not permeable and cause floods during rainy season. Infiltration of water is affected by soil erosion. The soil fertility is affected and it becomes saline because of over irrigation or water logging. Construction of dams and generation of hydro power changes the hydrological cycle. The problems of water crisis are:

1. **Quantity of Water:** Water deficit is a consequence of change in the water balance which is due to failure of rainfall. Precipitation by cloud seeding has achieved limited success. In many countries the major problem is how to assess and to provide water for the needs of the growing population.

2. **Quality of Water:** The use of water leads to the modification in physical, chemical and biological characteristics. This diminishes its quality and the water gets polluted. The water in almost all rivers is polluted and is not potable. Water problem is found everywhere: at home, in the cities and between the states and between the countries.

To solve the water problem proper planning is necessary. There is a need for estimation of water resources. Purity of water is a major water problem. Polluted or saline water is not useful even though the quantity is more. Desalinization of ocean water may bring change in the water cycle. To solve the water crisis the following steps should be adopted.

1. Using water economically
2. Recycle and reuse of water
3. Rainwater harvesting

1.2.3. Mineral Resources

Among natural resources, minerals occupy a distinct position, providing not only metals and fuels but also necessary raw materials for various types of industries and therefore they are of strategic importance in planning and development.

Minerals are non-renewable resources. They can be classified into three categories. They are metallic minerals, non-metallic minerals and fuel minerals. The common mineral resources are coal, iron-ore, manganese ore, mica, gold, bauxite, chromite, sulphur, copper, tin, nickel, lead, zinc, graphite, cobalt, atomic energy minerals and liquid fuels. India is very rich in mineral resources and there has been considerable exploitation of such resources for industrial advancement. Fuels (i.e., coal, lignite, petroleum and natural gas) constitute about 77% of the total mineral resources. The share of metallic and non-metallic minerals is about 15% and 8% respectively.

Environmental Effect of Extracting Minerals

The prosperous nations have largely raised the value of these mineral resources by their proper extraction and application in the industries. But no attention is being focused to the fact that in due course of time all the minerals will get depleted and as their quantity goes on decreasing, their cost of extraction will also go on increasing with the result that base of economic foundation will finally collapse. According to calculations by competent geologists, coal which is the conventional fossil fuel will hardly last for 300-400 years from now and petroleum and natural gas will be exhausted even earlier. Mining contaminates air, water and soil. Extraction of minerals also involves lot of occupational hazards and risks.

Minerals can be conserved through the following measures:

- By placing restrictions on reckless or over-exploitation.
- By reducing their waste through recycling.

- Efficient recovery of minerals from ores.

- Search for new deposits.

- Creation of Green belts in mining areas.

- Adopting safe mining processes in mining areas.

1.2.4. *Energy Resources*

Energy resources enrich the prosperity of life. Some of them that exist infinitely and never run out are called as Renewable. The rest that are in finite amounts and that took millions of years to form and will run out one day are called as Non-Renewable.

1. **Non-conventional or Renewable Resources:** Renewable energy sources are: available naturally in large quantities. The renewable resources are water, wind, sun and biomass.

Renewable energy resources are always available and will not run out due to tapping. These sources are carbon neutral. This means they do not produce Carbon compounds (such as other greenhouse gases). They do not pollute the environment. Renewable energy can be converted into electricity.

Renewable energy sources are:

Solar Energy: Solar energy is one of the most resourceful sources of energy. The total energy received each year from the sun is around 35,000 times the total energy used by man. However, about one - third of this energy is either absorbed by the outer atmosphere or reflected back into space. Solar energy is presently being used on a smaller scale in furnaces for homes and to heat up water. On a large-scale use, solar energy could be used to run cars, power plants and space ships.

Wind Energy: Wind power is another alternative energy source that could be used without producing by-products that are harmful to nature. Like solar power, harnessing the wind is highly dependent upon weather and location. The average wind velocity of earth is around 9m/sec. The power that could be produced when a wind mill is facing the wind of 10 mi/hr is around 50 watts. The moving air (wind) has huge amounts of kinetic energy, and this can be transferred into electrical energy using wind turbines. The wind turns the blades, which spin a shaft, which connects to a generator and makes electricity. The electricity is sent through transmission and distribution lines to a substation and then for use.

Hydroelectric Power: Water energy is the most conventional renewable energy source and is obtained from water flow and water falling from a height. The water under high pressure flows through the base of the dam and drives turbo-generators producing hydroelectric power. Hydro power is a clean, non-polluting source of energy. Hydropower has a low operating cost, once installed and can be highly automated.

Geothermal Energy: Geothermal energy is an alternative energy source, although it is not resourceful enough to replace energy needs of the future. Deep down in the earth's crust, there is molten rock (magma). Molten rock is simply rocks that have melted into liquid form as a result of extreme heat under the earth. This can be found about 1800 miles deep below the surface, but closer to the surface, the rocks layers are hot enough to keep water and air spaces there at a temperature of about 50-60°F. Geothermal power is obtained from the internal heat of the planet and can be used to generate steam to run steam turbine. This in turn generates electricity, which is a very useful form of energy.

Biomass Energy: Biomass is the term used to describe all the organic matter, produced by photosynthesis that exists on the earth's surface. The source of all energy in biomass is the sun. The biomass is acting as a kind of chemical energy store. Biomass is constantly undergoing a complex series of physical and chemical transformations and being regenerated while giving off energy in the form of heat to the atmosphere. Sophisticated technologies exist for extracting this energy and converting it into useful heat or power in an efficient way.

2. **Conventional or Non-renewable Resources:** These resources are limited in quantity. They will disappear in due course of time. Non-renewable energy is energy from fossil fuels (coal, crude oil, natural gas) and uranium. Fossil fuels are mainly made up of carbon. Non-renewable energy resources are:

Coal: When Coal is milled to a fine powder and allowed to burn more quickly, the hot gases and heat energy produced, converts water into steam. The high-pressure steam is passed into a turbine, rotates the turbine shaft and electricity is generated. As coal is obtained from mines safety measures are essential. In India the lignite availableis estimated as 34,168 million tonnes and from Neyveli alone we get 3,000 million tonnes.

Petrol: Petroleum is obtained from earth and under sea. It is obtained in crude form. It is purified to get Petrol, Kerosene, Diesel etc. Petroleum currently provides 90% of energy used for transportation.

Uranium: Uranium is a heavy metal naturally occurring in rocks, soil and even in ocean. Power generated from nuclear reaction is similar to fossil fuels. Energy from Uranium is called as Nuclear energy and the power generated from nuclear reaction is similar to that of fossil fuels.

Natural Gas: Natural Gas is colorless and odourless gas in its pure form. Unlike other fossil fuels, natural gas is clean while burning and emits lower levels of potentially harmful by products into the air. It is therefore called "Clean Gas'. Natural gas is used for cooking, transportation and in industries.

1.2.5. Land Resources

Land is an important natural resource and it is one of the most important components of the life supporting system but has been over-used and abused. The houses, roads and factories occupy nearly one third of the land. The forests occupy another one third of the land. The rest of land is farms, meadows and pastures. The soil forms the surface layer of land which covers more than the 80 percent of the land. The soil is defined as a natural body which keeps on changing and allows the plants to grow. It is made up of organic and in organic materials. In many countries, the cultivatable land area is shrinking continuously because of the growing population, poorly planned urbanization, deforestation, unscientific and indiscriminate construction of highways, airports and housing estates and conversion of agricultural land for industrial purposes. All these invariably lead to loss of land available for agricultural activities. Every year we are losing about 5 to 7 million hectares of agricultural land due to several natural and artificial (manmade) factors. This shows the seriousness of land degradation.

Causes of Degradation: Land degradation is any change in the condition of the land which reduces its productive potential. The factors responsible are:

1. **Water Erosion:** Out of 69 mha estimated to be critically degraded in India approximately 43 mha are non-arable and infertile, including 4 mha of ravine lands. The Himalayan Mountains with weak geological formation and poor physiographic conditions are under great stress and suffer from serious water erosion, though water erosion is also rampant in the Western Ghats and other areas of high intensity rainfall. Water erosion not only removes the productive surface layer of soil but also reduces the storage capacity of reservoirs.

2. **Wind Erosion:** Erosion by wind results in arid and in semiarid areas. Due to storm and high wind velocity, fine sand and soil particles are carried away from one place to another and there is formation of sand dunes. Soil erosion is affecting 15% of the earth's cropland area. Wind erosion is more prominent in the hot arid region occupying 31.7 mha of which 61 percent is found in western Rajasthan. Removal of vegetative cover and overgrazing enhance the intensity and extent of wind erosion and desertification.

3. **Water Logging:** Water logging caused by rise in water table poses a great threat to soil productivity. Roughly an area of 100,000 ha is estimated to be affected by water logging annually. Introduction of canal irrigation is the major reason for the fertile lands to be affected by water logging (e.g., Hissar, Haryana).

4. **Salinization and Sodification:** The development of soil salinity in India started long back and is more prominent in the arid and semi-arid areas. Continuous use of poor quality ground water for irrigation also leads to the development of soil salinity or sodicity, particularly in the slowly permeable soils. It is more serious in the Indo Gangetic Plain, black soil region, arid areas of Rajasthan and Gujarat and coastal areas. The saline soils have high concentration of chlorides and sulphates, lower values of pH and exchangeable sodium, better physical conditions etc. Many saline soils are often associated with high water table of poor groundwater quality.

5. **Nutrient Loss:** The transformation from high internal input agriculture in the past to the present day high external input (fertilizers, pesticides) agriculture causes this problem. Here the removal of plant nutrients is higher than what is added through the fertilizers.

6. **Soil Degradation by Pollution:** Soil degradation can be caused by pollution. Fertility of the soil can be severely damaged by pollution, even if there is no erosion. Soil is repository for all kinds of waste including hazardous and toxic waste. Spreading of sewage sludge on soil is harmful, since sewage contains heavy metals such as copper, cadmium and zinc. Sewage also contains large quantities of nitrates and phosphates but excessive use of these leads to soil degradation and deterioration in fresh water quality.

Desertification

Deserts are one of the important components of the Earth's environment. They are fragile ecosystems where vegetation is absent or scarce as moisture is insufficient to support it. Desertification is defined as land degradation in the arid, semi-arid and dry sub-humid areas, due to various factors such as climatic variations and human activities. It ultimately leads to a reduction or loss of production potential of the land and is a major environmental problem in the Asian countries. The net result is desertification/degradation of more land, as well as a gradual deterioration of the production potentials of these climatically handicapped fragile regions.

Deforestation, soil erosion, urbanization, pollution etc. led to desertification. To meet the energy requirements for increased population, new power stations, dams, reservoirs etc. are constructed which leads to deforestation. Trees are destroyed for meeting increased fuel requirements. Deforestation to meet energy requirements as well as to provide new agricultural land compensating for the land loss in urbanization process results in deserts. One way or the other 80% of the productive land in the arid and semiarid areas of the globe is affected by desertification.

Impact of Land Degradation on Environment

The immediate impact of land degradation includes reduced crop yields, increasing need for agricultural inputs and decreasing profits, reduction in the value of land, loss of water resources due to frequent and severity of floods, siltation of reservoirs, rivers etc., and adverse effect on power generation. Standard of living, which depends upon farm income greatly, declines with decrease in the per capita availability of land. Increasingly less remunerative subsistence farming systems lead the peasants to leave the land uncultivated which then tends to become barren. All the above said reasons ultimately result in shifting the land resources from agriculture to other uses, which bring environmental degradation. The land degradation can be overcome through the following strategies:

1. Proper records of land productivity status need to be prepared with the help of the soil scientists along with the latest technologies like remote sensing.
2. For each soil property there should be a clear demarcation of boundary inorder to monitor soil, which is stressed by external degrading agencies.
3. Industrial activities in the nearby areas which are hazardous to the environment, need to be regulated through laws.

4. Sustainable and effective land use system should be followed.

5. Implementation of agro-forest ecosystem wherever water erosion is a serious threat.

6. Adoption of integrated watershed management system and integrated balanced nutrient management system.

1.2.6. Food Resources

Food is any substance consumed to provide nutritional support for the body. It is usually of plant or animal origin, and contains essential nutrients, such as carbohydrates, fats, proteins, vitamins or minerals. The substance is ingested by an organism and assimilated by the organism's cells in an effort to produce energy, maintain life and to stimulate growth. Historically, people secured food through two methods namely, hunting and agriculture. Today, most of the food energy consumed by the world population is supplied by the food industry, which is operated by multinational corporations that use intensive farming and industrial agriculture to maximize output.

Sources of Food Supply

The following are the important sources of food supply.

1. Agriculture

Agriculture is the cultivation of plants, fungi and other life forms for food, fibre and other products used to sustain life. This is the basic source of food supply. Agriculture is the key factor in the rise of sedentary human civilization, whereby farming of domesticated species created food surpluses that nurtured the development of civilization.

Traditional agriculture usually involves a small plot, simple tools, naturally available water, organic fertilizers and mix of crops. Modern agriculture makes use of hybrid seeds of selected and single crop variety, high-tech equipments, chemical fertilizers, pesticides, insecticides and modern irrigation methods to obtain increased production.

2. Sea Food

Sea food refers to any sea animal or plant that is served as food and eaten by humans. Sea food includes sea water animals such as fish, shellfish etc. The harvesting of wild sea food is known as fishing and the cultivation and farming of sea food is known as aquaculture. Sea food is an important source of protein in many diets around the world, especially in coastal areas. Studies on sea food revealed that various species of sea food pointing to a collapse due to pollution and over fishing, threatening oceanic ecosystems.

3. Livestock

Livestock refers to one or more domesticated animals raised in an agricultural setting to get food, fiber etc. Raising animals is an important component of agriculture. It has been practiced in many cultures since the transition to farming from hunter-gather lifestyles.

World Food Supply

In the ancient period, the world supply of food was very limited, because, the main sources of food were hunting and pastoral farming. Later on, agricultural revolution increased the world food supply. During this period, people adopted agriculture as a main source of food. But the world supply of food was not adequate to support the growing population.

During the second stage of agricultural revolution, with the application of science and technology, the world production and supply of food increased remarkably. The adoption of multiple cropping, mixed farming, use of irrigation for cultivating the dry soils, control of soil erosion and floods, use of quality seeds, use of chemical fertilizers and green manure, use of pesticides and insecticides, availability of storage and marketing facilities, availability of finance for agricultural activities, remunerative prices for food crops, etc. Have contributed to increase in food supply. Later on the green revolution that has taken place in many countries including India resulted in remarkable increase in food production.

Effects of Modern Agricultural Practices on Environment

Modern agricultural practices have substantially changed the farming, crop production and harvesting. On the other hand it leads to several ill effects on environment.

Some local and regional changes of modern agricultural practices are:

1. It leads to soil erosion.

2. It results in the increase in sedimentation towards downstream of the river.

3. Alteration in the fertility of soil.

4. Increase in deforestation due to more land under cultivation.

5. Leads to soil pollution.

6. It leads to desertification.

7. It results in the change in the ecology of estuaries due to increase in sedimentation at the junctions of rivers.

Effects of Artificial Chemical Fertilizers

1. Increase in water borne diseases due to contamination of surface and ground water resources.
2. Loss of natural fertility of the soil.
3. Loss of organic matter from the soil.
4. Threat to the quality of drinking water due to disposal of fertilizers into landfills, sites and lands.

Effects of Pesticides

Pesticides are the chemicals used in the soil to kill pests. Following are its disadvantages:

1. Species which are not targeted are also killed or injured.
2. After sometime the pest develop resistance against the pesticides.
3. Soil fertility is reduced.
4. On short duration exposure it causes illness and slow poisoning to human beings.
5. On long duration exposure it causes cancer, genetic defects, immunological and other chronic diseases.

1.3. Role of an Individual in the Conservation of Natural Resources

The survival and wellbeing of a nation depend on its sustenance or sustainable development. It is a process of social and economic betterment that satisfies the needs and values of all. To this end, every man must ensure that the demand on the environment does not exceed its carrying capacity for the present as well as for future generations.

Over the years, there has been progressive pressure on the environment and the natural resources, the alarming consequences of which are becoming evident in increasing proportions. These consequences detract from the gains of development and worsen the standard of living of the poor who are directly dependent on natural resources. It is in this context that there is a need to give a new thrust towards conservation and sustainable development.

Conservation of Natural Resources

Man is an integral part of the environment, exchanging materials with the environment in a continuous cycle and has the ability to modify the environment using his knowledge. As population increases more space is required and resources are used exhaustively. But it will affect the present as well as future generation.

It must, therefore be realized that man should try to conserve resources. So, there is a need to educate the people about their role in conserving natural resources and safeguarding the environment.

Environmental education must first and foremost create awareness and sensitivity to environment and allied problems. Environmental education must start at home and its neighborhood by developing manipulative skills in home activities and play, personal, hygiene and problems of food and water contamination. The entire process of environmental education should be gradual right from childhood and it will be a lifelong process, as environment is part of our existence and life.

Individual's Contribution for Conservation of Natural Resources

a. *Conservation of Water*

1. Using minimum water for all domestic purposes.
2. Checking the water leaks and repairs properly.
3. Reusing the soapy water, after washing clothes for washing courtyard, carpets etc.
4. Using drip irrigation in agriculture.
5. Installation of rain water harvesting system in all houses.
6. Installation of sewage treatment plant in industries and institutions.

b. *Conservation of Soil*

1. Growing different types of plants i.e. trees, herbs and shrubs.
2. Avoiding strong flow of water during irrigation.
3. Soil erosion can be prevented by sprinkler irrigation.
4. Using green manures.
5. Adopting mixed cropping.

c. *Conservation of Energy*

1. Economical use of electrical energy.
2. Using solar power for domestic purposes.
3. Riding bicycle or walking instead of using two- wheelers for a short distance.

d. *Conservation of Food Resources*

Food processing and food preservation techniques to be adopted especially when there is over production.

e. ***Conservation of Forests***

1. Using non timber product.
2. Planting more trees.
3. Grazing must be controlled.
4. Minimizing the use of paper and fuel.
5. Avoiding the construction of roads in the forest areas.

1.4. Conclusion

Human beings have to realize the fact that there is need to judiciously utilize the available resources to the best of their advantage so as to prevent the destruction of these available resources. It must also be emphasized that any imbalance in the utilization of natural resources will result in an irreversible change to the quality of the life-supporting environment.

CHAPTER II

ECOSYSTEM

An ecosystem is a combination of two words: "ecological" and "system." Together they describe the collection of biotic and abiotic components and processes that comprise and govern the behavior of some defined subset of the biosphere. An ecosystem is a natural unit consisting of all plants, animals and micro-organisms (biotic factors) in an area functioning together with all of the non-living physical (abiotic) factors of the environment. Ecosystems vary in size. Some ecosystems are very large, some may be physically small, such as a meadow at the edge of a forest, or a coral reef in the ocean.

Living organisms cannot live isolated from their non-living environment because the latter provides materials and energy for the survival of the former i.e. there is an interaction between a biotic community and its environment to produce a stable system-a natural self-sufficient unit which is known as an ecosystem. Within each ecosystem, there are habitats, a place where a population lives. A population is a group of living organisms of the same kind living in the same place at the same time which interact to form a community. The habitat must supply the needs of organisms, such as food, water, oxygen, and minerals, if not the organisms will move to a better habitat. Two different populations cannot occupy the same niche at the same time. So the processes of competition, predation, cooperation, and symbiosis occur.

An ecosystem is, therefore, defined as a natural functional ecological unit comprising of living organisms (biotic community) and their non-living (abiotic or physicochemical) environment that interact to form a stable self-supporting system. Pond, lake, desert, grassland, meadow, forest etc. are common examples of ecosystems.

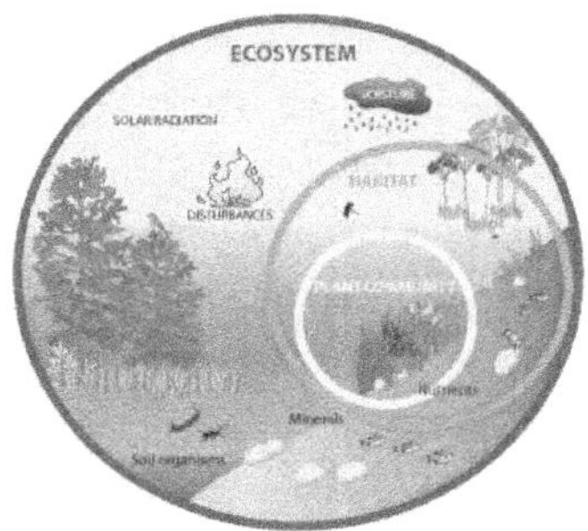

Eco System

"

2.1. Structure of an Ecosystem

The ecosystem consists of two components.

a. Biotic Component or living component.

b. Abiotic Component or non-living component.

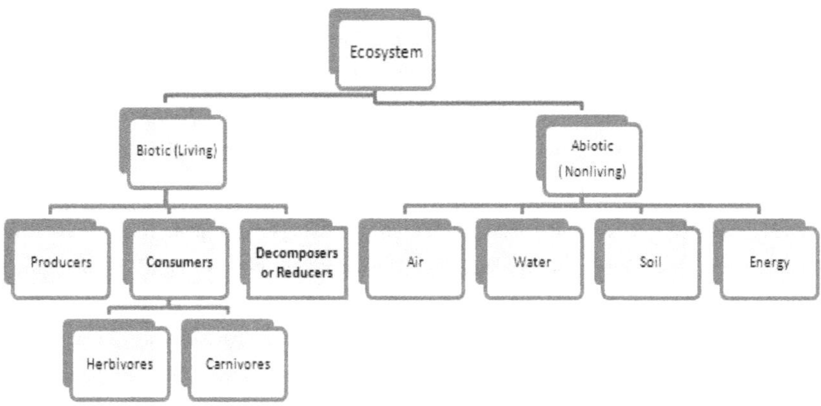

Structure of an Ecosystem

A. *Biotic Components*

The living organisms including plants, animals and microorganisms (Bacteria and fungi) that are present in an ecosystem form the biotic components. On the basis of their role in the ecosystem the biotic component can be classified into three main groups namely Producers, Consumers and Decomposers.

1. **Producers:** The green plants have chlorophyll with the help of which they trap solar energy and change it into chemical energy of carbohydrates using water and carbon dioxide. This process is known as photosynthesis. As the green plants manufacture their own food they are known as Autotrophs (i.e. auto = self, trophos = feeder). The chemical energy stored by the producers is utilised partly by the producers for their own growth and survival and the remaining is stored in the plant parts for their future use.

2. **Consumers:** The animals lack chlorophyll and are unable to produce their own food. Therefore, they depend on the producers for their food. They are known as heterotrophs (i.e. heteros = other, trophos = feeder).

3. **Decomposers or Reducers:** Bacteria and fungi belong to this category. They breakdown the dead organic materials of producers (plants) and consumers (animals) for their food and release to the environment the simple inorganic and organic substances produced as by-products of their metabolisms. These simple substances are reused by the producers resulting in a cyclic exchange of materials between the biotic community and the abiotic environment of the ecosystem. The decomposers are known as Saprotrophs (i.e., sapros = rotten, trophos = feeder).

B. *Abiotic Components*

The non-living factors or the physical environment prevailing in an ecosystem form the abiotic components. They have a strong influence on the structure, distribution, behavior and inter-relationship of organisms.

The list of abiotic factors are clouds, rainfall, altitude, temperature, oxygen, salinity, soil (edaphic factors), air, water, sunlight, humidity, topography, pH, atmospheric gases etc.

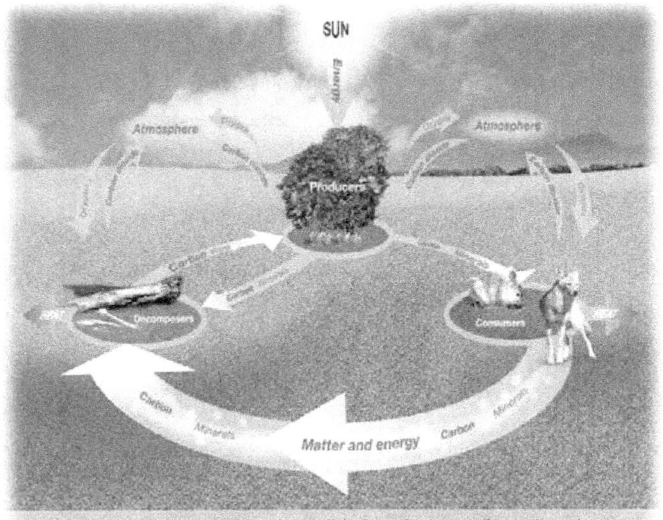

Living and Nonliving Parts of an Ecosystem

2.2.   Functions of an Ecosystem

Ecosystem is a functional unit, consisting of all living organisms, such as plants, animals and microbes in a given area, and all non-living physical and chemical factors of environment, linked together through nutrient cycling and energy flow. The functioning of an ecosystem is important because all its components are dynamic and are responsible for the creation of the unique state of man-environment relationship.

The functioning of an ecosystem consists of,

a)   Energy Flow

b)   Productivity

c)   Bio-geochemical cycling

A.   *Energy Flow in the Ecosystem*

Energy is the ability to do work or to move matters. The movement of energy through the ecosystem is called energy flow. The energy flow is unidirectional in an ecosystem and flows from producers to herbivores then to carnivores. It cannot occur in the reverse direction.

The sun is the main source of energy for the environment. The radiant energy from the sun travels through the space in the form of waves. But only a small fraction of solar radiation reaches the earth to provide energy for the biotic components of the ecosystem. The energy reaching the earth's surface is used by the green plants and other organisms during photosynthesis by converting the light energy to chemical energy and making it available to other organisms as food.

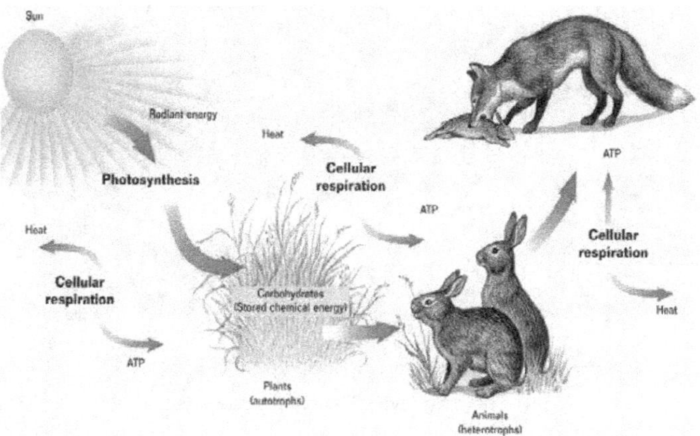

B. Productivity

Productivity of an ecosystem means the rate of production, i.e., amount of organic matter produced or accumulated by plants or the producers per unit time and area. It is of three types.

- Primary Productivity
- Secondary Productivity
- Net Productivity

i) Primary Productivity

Primary Productivity refers to the rate at which radiant energy is stored by photosynthetic and chemosynthetic activities of producer organisms, mainly green plants in the form of organic substances which can be used as food material. Primary productivity is of two types. They are Gross Primary Productivity which is the total rate of photosynthesis including organic matter used up in respiration during a particular period and Net Primary Productivity which is the amount of organic matter stored in plant tissues in excess of that used up by the plants during respiration.

ii) Secondary Productivity

Secondary productivity is the rate of energy storage at consumer level. It actually remains mobile from one organism to another organism and does not live in situ like primary productivity.

iii) Net Productivity

Net productivity refers to the rate of storage of organic matter not used by the consumers at the consumer level or second trophic level. It is the rate of increase of biomass of the primary producers, which has been left out by the consumers.

C. Bio-Geochemical Cycles

Primary producers take up the elements from the ecosystem to build their biomass which is utilized by consumers. After the death of producers and consumers the microbes decompose the dead organisms and release the chemical nutrients back to the environment. These minerals are absorbed by the green plants through their roots. Thus, the cycle of minerals or nutrients is completed.

Chemicals available in nature cannot be used in that form unless these chemicals undergo the cycle in complex paths through the living and non-living components of the ecosystem. By biological, geological and chemical processes, these naturally available chemicals are converted into useful forms. Such processes are called biogeochemical cycles.

There are many biogeochemical cycles in nature through which the ecosystem functions. They are:

i) Hydrological Cycle or Water Cycle

Hydrological cycle or water cycle involves interchange of water between the earth's surface and the atmosphere through rainfall and evaporation. The water from water bodies like oceans, seas, rivers, lakes, etc. Gets evaporated by solar energy. These water vapors, after cooling and condensation, form clouds, and resulted in rainfall, snowfall, etc. A large part of the rainfall occurs in ocean and seas. A sizeable part of water goes back to the ocean and seas through rivers and streams. Some part of water infiltrates into the soil and becomes underground water. A small quantity of water is absorbed by the plants and other animals, and the same is released during respiration and transpiration of plants. Thus, there is a continuous cycling of water. Organisms play an important role in water cycle. Most organisms contain a significant amount of water i.e., upto 90% of their body weights.

ii) Carbon Cycle

The small amount of carbon dioxide in the atmosphere is the only source of carbon that passes through the food chain. Carbon dioxide moves from the atmosphere to the green plants (producers), then to animals (consumers) and finally to bacteria and other microorganism (decomposer) that return into the atmosphere through decomposition of dead organic matter.

iii) Oxygen Cycle

Oxygen constitutes about 21% of the atmospheric air. The plants release oxygen during photosynthesis. Gaseous oxygen is used by all the organisms for oxidation of the organic matter-a process known as respiration.

iv) Nitrogen Cycle

Nitrogen is an essential element of all forms of life. The chief source of nitrogen is the atmosphere. Atmosphere has about 79% nitrogen. Nitrogen is never taken directly from the atmosphere. The chief source of nitrogen for plants is nitrate in the soil. Nitrogen is converted into usable form by nitrogen fixing bacteria.

v) Phosphorus Cycle

Phosphorus cycle is a sedimentary cycle. The great reservoir of phosphorus are rocks or other deposits which have been formed in the past geological ages. The rocks and other deposits erode and release phosphates to ecosystems. The phosphates absorbed by plants pass through the food chain.

2.3.   Food Chain and Food Web

2.3.1. *Food Chain*

In an ecosystem energy flows from one trophic level to another. A trophic level represents a group of organisms, which are either predators or prey.

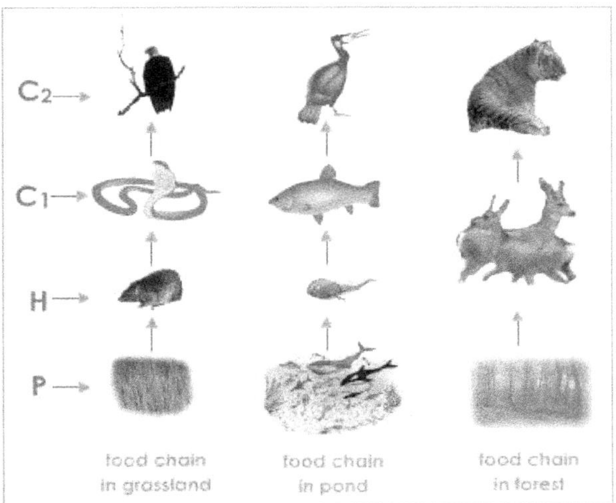

All organisms in an ecosystem are linked to one another for their nutritional needs. This relation between the individuals constitutes a linear chain called the food chain. Food chain shows the relationship between producers, consumers, and decomposers, showing who eats whom. Food chain shows how each member of an ecosystem gets its food. A simple food chain links a producer, a herbivore, and one or more carnivores. Arrows in the food chain show how energy is passed from one link to another.

Producers → Herbivores → Carnivores

Food chain is of two kinds namely Grazing food chain and Detritus food chain. The grazing food chain starts from green plants passes on to herbivorous primary consumers and ends with carnivorous animals. Detritus or decomposing food chain begins with dead organic matter goes to microorganisms and then passes on to organisms that feed on derivers (organisms that - eat detritus) and their predators. Ecosystems of this type are less dependent on direct solar energy. On the other hand they depend on the supply of organic matter produced by another ecosystem.

2.3.2. *Food Web*

In an ecosystem, the relationships between the food chains are inter-connected. These relationships are very complex, as one organism may be a part of multiple food chains. Hence, a web like structure is formed in place of a linear food chain. A food web is a graphical depiction of feeding connections among species of an ecological community. The food web also defines the energy flow through the species of a community as a result of their feeding relationships. Thus, a food web consists of many interconnected food chains in which most of the communities include various populations of producer organisms which are eaten by any number of consumer populations. The green crab, for example, is a consumer as well as a decomposer. The crab will eat dead things or living things if it can catch them. A secondary consumer may also eat any number of primary consumers or producers. In a food web, nutrients are recycled in the end by decomposers. Animals like shrimp and crabs can break the materials down to detritus. Then bacteria reduce the detritus to nutrients. Decomposers work at every level, setting free nutrients that form an essential part of the total food web.

2.3.3. *Significance of Food Chain and Food Web*

Food chains and food webs play a very important role in the ecosystem, because the most important functions of energy flow and nutrient cycling take place through food chain and food web. Food chain also helps in maintaining and regulating the population size of different animals and helps to maintain ecological balance. Study of food chain helps in understanding the feeding relationships and the interactions between organisms in any ecosystem. Food web also provides information about the biological diversity of ecosystem.

2.4.   Ecological Pyramids

The graphical representation of the trophic structure and trophic function is called an ecological pyramid. The concepts of the ecological pyramid were developed by Charles Elton and are called as Etonian Pyramids.

There are three types of ecological pyramids.

    a)   Pyramid of numbers

    b)   Pyramid of biomass

    c)   Pyramid of energy

A. *Pyramid of Numbers*

A Pyramid of Numbers can be generated by counting all the organisms at the different feeding levels. The number of organisms gradually decreases from the base to the top of the pyramid in the eco system from producers to secondary consumers. Thus, the decrease in number occurs because of the energy loss, when one organism feeds on another organism. Depending upon the type of ecosystem and food chains, the pyramid of number may be upright or inverted.

In a grass land ecosystem, the producers are mainly the grasses and the grasses are always maximum in number. The primary consumers are herbivores like rabbits, mice etc., and they are lesser in number than the producers i.e., the plants. The secondary consumers are carnivores like snakes and lizards and they are lesser in number than the primary consumers. The tertiary consumers are the top carnivores like hawk or birds and they are least in number in a grassland ecosystem and the pyramid of numbers becomes upright.

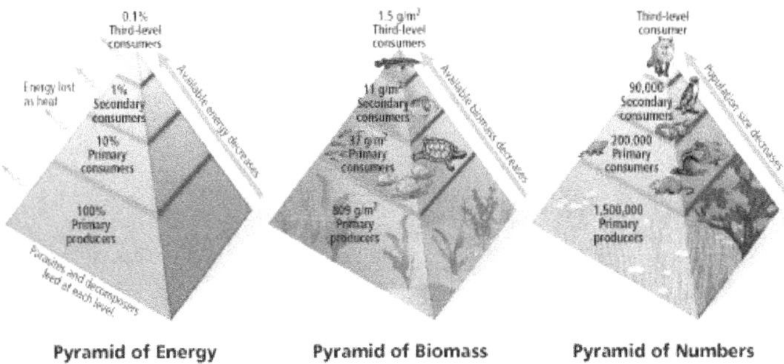

Pyramid of Energy   **Pyramid of Biomass**   **Pyramid of Numbers**

However, in the parasitic food chain the pyramid is always inverted. The number of organisms gradually increases making the pyramid inverted in shape.

B. *Pyramid of Biomass*

The organisms are collected from each feeding level, dried and weighed. This bio mass (dry weight) represents the amount of available energy of the organisms. The total biomass of successive trophic level is arranged to form a pyramid, and that is called Pyramid of biomass. The Pyramid of biomass may be upright or inverted. In a grass land ecosystem or a forest ecosystem, there is generally a gradual decrease in the biomass of organisms at successive trophic levels from producers to the tertiary consumers or top carnivores as a result, the Pyramid of biomass become upright.

C. Pyramid of Energy

The Pyramid of energy deals with the relationship of energy accumulation pattern at different trophic levels of a food chain. It explains about the productivity and energy flow of the ecosystem. The Pyramid of energy is always upright because there is a gradual decrease in the energy content at successive trophic levels from the producers to consumers. The primary producers of an ecosystem trap the radiant energy of the sun, and convert it into chemical energy. The trapped energy, flowing in the food chain from the producers to the tertiary consumers or the top carnivores, decrease at successive trophic levels, as a result, the pyramid of energy is always upright.

2.5. Types of Ecosystems

There are two primary types of ecosystems namely Natural Eco System and Artificial Eco System. Natural ecosystems may be terrestrial (meaning desert, forest, or meadow) or aquatic, (pond, river, or lake). A natural ecosystem is a biological environment that is found in nature (e.g. a forest) rather than created or altered by man (a farm). Humans have modified some ecosystems for their own benefit. These are artificial ecosystems. They can be terrestrial (crop fields and gardens) or aquatic (aquariums, dams, and manmade ponds).

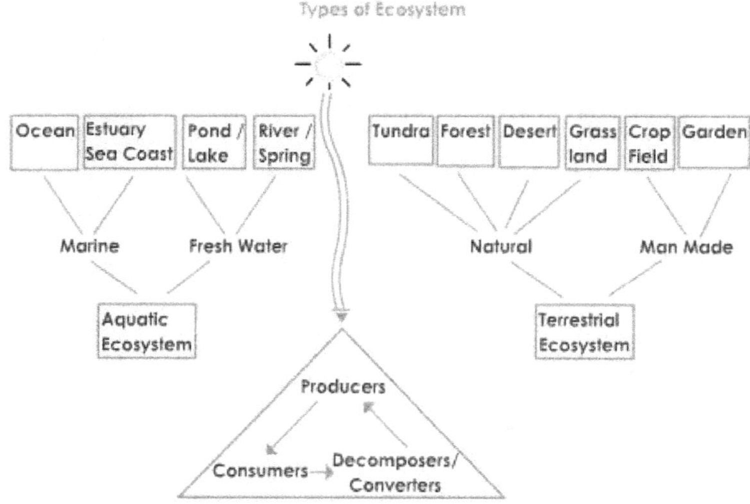

In general, there are two kinds of ecosystems: Terrestrial and Aquatic. Any other sub-ecosystem falls under one of these two systems.

A. Terrestrial Ecosystems

Terrestrial ecosystems are found everywhere apart from water bodies. They are broadly classified into Forest Ecosystem, Grassland ecosystem and Desert ecosystem.

i) Forest Ecosystem

The Forest ecosystems are areas of the landscape that are dominated by trees and consist of biologically integrated communities of plants, animals and microbes, together with the local soil (substrates) and atmospheres (climates) with which they interact. They are the ecosystems in which an abundance of flora or plants occur. Therefore, in forest ecosystems the density of living organisms is quite high. A small change in this ecosystem could affect the whole balance, effectively bringing down the whole ecosystem. Forest Ecosystem is divided into:

Tropical Deciduous Forests: Tropical deciduous forest is found in quite a few parts of the world where a large variety of fauna and flora are found. It consists of shrubs and dense bushes along with a broad collection of trees.

Temperate Deciduous Forests: The forest is located in the moist temperate places that have sufficient rainfall. The trees shed the leaves during the winter months.

Taiga: Situated just before the arctic regions, the taiga is defined as evergreen conifers. As the temperature is below zero for almost half a year, for the remainder of the months, it buzzes with migratory birds and insects.

ii) Grassland Ecosystem

Grasslands are natural ecological communities dominated by grasses and with less than 10% natural tree or shrub cover. They contain many grass species and greater diversity of other herbs. Grasslands are located in both the tropical and temperate regions of the world though the ecosystems vary slightly. The main vegetation includes grasses legumes and plants of the composite family. A lot of grazing animals, insectivores and herbivores inhabit the grasslands. Grasslands are known by various names in different parts of the world as listed below.

Place	Name of the grassland
North America	Prairies
Eurasia (Europe and Asia)	Steppes
Africa	Savanna
South America	Pampas
India	Savanna

iii) Desert Ecosystem

Desert ecosystems are located in regions that receive an annual rainfall less than 25 percent. Due to the extremely high temperature, low water availability and intense sunlight, fauna and flora are scarce. Vegetation is mainly shrubs, bushes, few grasses and rarely trees. The stems and leaves of the plants are modified in order to conserve water as much as possible. The best known desert plants are the succulents such as the spiny leaved cacti. The fauna includes insects, birds, camels, reptiles all of which are adapted to the desert (xeric) conditions.

Characteristics of Desert Ecosystem

Desert plants are adapted to hot and dry conditions and they are:

- Mostly shrubs.
- Leaves absent or reduced in size.
- Leaves and stem are succulent and water storing.
- In some plants even the stem contains chlorophyll for photosynthesis.
- Root system spread over large area.

Desert animals are physiologically and behaviorally adapted to desert conditions and they are:

- Fast runners.
- Nocturnal in habit to avoid the sun's heat during day time.
- Conserve water by excreting concentrated urine.
- Animals and birds usually have long legs to keep the body away from the hot ground.
- Herbivorous animals get sufficient water from the succulents they feed.

B. Aquatic Ecosystems

An aquatic ecosystem is an ecosystem found in a body of water. It encompasses aquatic flora, fauna and water properties, as well. Aquatic ecosystems are critical components of the global environment. In addition to being essential contributors to biodiversity and ecological productivity, they also provide a variety of services for human populations, including water for drinking and irrigation, recreational opportunities, and habitat for economically important fisheries. However, aquatic systems have been increasingly threatened directly and indirectly by human activities.

On the basis of salinity, Aquatic ecosystems are classified into two types namely Fresh water eco system and Marine eco system.

i) *Fresh Water Ecosystem*

Water on land which is continuously cycling and has low salt content is known as fresh water and its study is called limnology. Sources of fresh water ecosystem are:

- Static or still water (Lentic) e.g. pond, lake.
- Running water (Lotic) e.g. springs, mountain brooks, streams and rivers.

Characteristics

- Fresh water has a low concentration of dissolved salts.
- The temperature shows diurnal and seasonal variations.
- In temperate regions, the surface layer of water freezes but the organisms survive below the frozen surface.
- Light has a great influence on fresh water ecosystems. A large number of suspended materials obstruct penetration of light in water.
- Certain animals float up to water surface to take up oxygen for respiration.
- Aquatic plants use carbon dioxide dissolved in water for photosynthesis.
- Lakes and ponds are inland depressions containing standing water.

ii) *Marine Ecosystem*

The marine ecosystem is the largest ecosystem on earth Oceans cover approximately 70% of the earth's surface with an average depth of 2.4 miles, or 3,800 meters. The marine ecosystem, in addition to the temperate and tropical oceans, includes the shorelines, with mud flats, rocky and sandy shores, tidepools, barrier islands, estuaries, salt marshes, and mangrove forests making up the shoreline segment. Marine ecosystems support a great diversity of life and variety of habitats. The ocean has a major influence on weather and climate.

Characteristics of Marine Ecosystem

Location: Conventionally, the ocean has been divided into four major ocean basins: Atlantic, Pacific, Indian and Arctic oceans. Specific marine ecosystems such as coral reefs, estuaries, salt marshes, and mangrove forests are found throughout the world, but are characteristic of certain areas, depending on climate, geography, water temperature, and other physical factors.

Plants: Marine habitats are the home to seaweeds, or marine algae (brown, green, red), sea grasses, which are the only marine flowering plants, and mangroves, located on muddy tropical shores.

Animals: Marine ecosystems are homes to protozoans, marine invertebrates (echinoderms, molluscs, segmented and non-segmented worms, jellies, coral, sea anemones, hyroids) marine vertebrates (fishes, birds, mammals), and plankton (phyto and zooplankton).

Climate: Monsoon, tropical, subtropical, temperate, polar, subpolar.

2.6. Conclusion

Ecosystem management is need of the hour to conserve major ecological systems and restore natural resources while meeting the socio economic, political and cultural needs of current and future generations.

CHAPTER III

BIO – DIVERSITY

Biodiversity refers to the variety of life. It is seen in the number of species in an ecosystem or on the entire Earth. Biodiversity is a complex and balanced network of different species which are mutually dependent upon each other.

Biodiversity can be defined as the variety and variability among living organisms and the ecological environment in which they exist. Biodiversity is closely related to the function and stability of communities and ecosystems. It involves all the species of plants, microorganisms and animals on earth.

Genetic variations among individuals of the same species, as well as among different species constitute the basis for biodiversity. The existing species of plants and animals are the products of alteration, recombination and natural selection of genes. Biodiversity is in a way a kind of bank for genes.

Living things, as they evolved on the earth, have produced a bewildering gene diversity. There are millions of species of plants and animals but even within a species, there exists enormous diversity. As new varieties of plants and animals useful to human beings are bred, this massive gene bank provided by nature becomes significant.

Biodiversity can be studied at three levels namely genes, species and ecosystems. Genes are the components of species, and species are the components of ecosystems, altering the make-up of any level of this hierarchy can change the others.

3.1.   Meaning of Biodiversity

Biodiversity means the variability among living organisms from all sources like terrestrial, marine and other aquatic ecosystems. The ecological complexities are due to the diversity at higher level.

Definition of Biodiversity

Reid and Miller (1989) defined biodiversity as: "the variety of the world's organisms, including their genetic diversity and the assemblage they form".

Levels of Biodiversity

Biodiversity can be generally described in terms of its three fundamental and hierarchically related levels of biological organization. They are:

1. Genetic biodiversity
2. Species biodiversity
3. Ecological biodiversity
    1. **Genetic Biodiversity** is a measure of the variety available for the same genes within individual species.
    2. **Species Diversity** describes the different kinds of organisms within individual communities or ecosystems.
    3. **Ecological Diversity** is the richness and complexity of a biological community, including number of trophic levels and ecological processes that capture energy sustain food webs and recycle materials within this system.

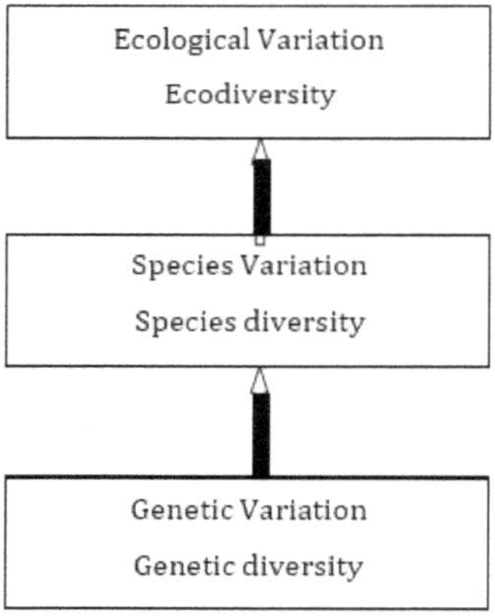

Levels of Bio Diversity

3.2. Functions of Biodiversity

There are two main functions of biodiversity. Firstly, biodiversity stabilizes the climate, water, soil, and the overall health of the biosphere. Secondly, biodiversity is the source of species on which the human race depends for food, fodder, fuel, fibre, shelter and medicine.

Biodiversity plays a role in regulating natural processes such as the growth cycles of plants and weather systems. The significance of biodiversity can be summarized as follows.

Food and Drink

Biodiversity provides food for humans. About 80% of our food supply comes from just 20 kinds of plants (Rice, Wheat, Cholam etc.). Although many kinds of animals are used as food, again most consumption is focused on a few species (Goats, Sheeps, Fish, Chicken etc). There is a vast untapped potential for increasing the range of food products suitable for human consumption.

Medicines

A significant proportion of drugs are derived directly or indirectly from biological sources. However, only a small proportion of the total diversity of plants has been thoroughly investigated for potential sources of new drugs.

Industrial Materials

A wide range of industrial materials are derived directly from biological resources.

These include timbers, fibers, dyes, resins, gums, adhesives, rubber and oil.

Ecological Services

Biodiversity plays a vital role in regulating the chemistry of our atmosphere and water supply. It is directly involved in recycling nutrients and providing fertile soils.

Leisure, Cultural and Aesthetic Values

Many people derive pleasure from biodiversity through leisure activities such as enjoying a walk in the countryside, bird watching and viewing programmed on nature in television. Biodiversity has inspired musicians, painters, sculptors, writers and other artists. Many cultural groups view themselves as an integral part of the natural world and show respect for other living organisms.

Because of these reasons many organizations like the World-Wide Fund for Nature (WWF) and the United Nations Commission on Environment and Development (UNCED) stressed the need to preserve biodiversity.

3.3.   Values of Biodiversity

India is one of the top 12 mega diversity countries. Diversity in genes, species and ecosystems provides the raw materials with which human communities adapt to change. Thus, the loss of each additional species, gene and ecosystem reduces the ability of nature and people to adapt to the changing environment.

Levels of Biodiversity

The value of biodiversity is classified into direct and indirect values. Biodiversity has direct consumptive value in agriculture, medicine and industry. The indirect value of biodiversity includes ecosystem's ability to absorb pollution, maintain soil fertility and microclimates, recharge ground water, and provide other invaluable services. The multiple uses of bio-diversity values are:

1. **Consumptive Use Value:** Consumptive use value refers to assessing the value of natural resources which are commercially harvested such as firewood, fodder, timber, fishery resource, minor forest products, animal meat and others. These are for direct consumption that generates economic benefits. Nearly 99% of the human food supply comes from the land. In India, the present per capita availability of land is less than 0.20 ha. Globally, about one third of agricultural land is devoted to crops and the remaining two-third is devoted to pastures for livestock grazing.

   a. **Food:** A large number of wild plants are consumed by human beings as food. About 90% of present-day food crops have been domesticated from wild tropical plants. A large number of wild animals are also our sources of food.

b. **Drugs and Medicines:** About 75% of the world's population depends upon plants or plant extracts for medicines. The wonder drug penicillin used as an antibiotic is derived from a fungus called Penicillium. Likewise, we get Tetracyclin from a bacterium. Quinine, the cure for malaria is obtained from the bark of Cinchona tree, while Digital in is obtained from foxglove, which is an effective cure for heart ailments. Vinblastine and vincristine, two anticancer drugs, have been obtained from periwinkle plant, which possesses anticancer alkaloids.

c. Our forests have been used since ages for fuel wood. The fossil fuels like coal, petroleum and natural gas are also products of fossilized bio-diversity.

2. **Productive Use Value:** Productive use values are the commercially usable values where the product is marketed and sold. These may include the animal products like tusks of elephants, musk from musk deer, silk from silk worm, wool from sheep, lac from insects, etc., all of which are traded in the market. Many industries are dependent upon the productive use values of biodiversity, example the paper and pulp industry, Plywood industry, Railway sleeper industry, Silk industry, Ivory works, Leather industry, Pearl industry etc.

3. **Social Value:** These are the values associated with the social life, customs and religion of the people. Many of the plants are considered holy and sacred in our country like Tulsi, Peepal, Mango, Lotus etc. The leaves, fruits or flowers of these plants are used in worship or the plant itself is worshipped. Many animals like Cow, Snake and peacock also have significant place in our Psycho–Spiritual spheres.

4. **Ethical Value:** Environmental ethics includes the relationship between man and environment in which people are living and the views of different religions on nature. Hinduism expresses the firm belief that the natural environment in which people are living is a manifestation of divine nature itself. Similarly various religions like Buddhism, Judaism, Christianity, Taoism, Islam have insisted that the mind is supreme and environment results from a proper moral and spiritual lifestyle.

It is also sometimes known as existence value. It involves ethical issues like "all life must be preserved". The ethical value means that we may or may not use a species, but knowing the very fact that the species exist in nature gives us pleasure. We are not deriving anything direct from Kangaroo, Zebra or Giraffe but we all strongly feel that these species should exist in nature.

5. **Aesthetic Value:** Every kind of animal and plant differs from each other and contributes in a special way to the beauty of nature. It also heightens the enjoyment of camping and other forms of outdoor recreation. Forests are not only valued for their products but they support butterflies, grasshoppers, beetles and other animal and plants that have aesthetic beauty.

6. **Option Value:** Option value is the national value (i.e) economic, cultural or scientific values of a Nation. These values include the potentials of biodiversity that are presently unknown and need to be explored. There is a possibility that we may have some potential cure for AIDS or cancer existing within the depths of a marine ecosystem, or a tropical rainforest. Thus option value is the value of knowing that there are biological resources existing on this biosphere that may one day prove to be an effective option for something important in the future.

7. **Ecological Value:** Ecological value refers to the services provided by ecosystems such as prevention of soil erosion, prevention of floods, maintenance of soil fertility, nutrients cycles, fixation of nitrogen, hydrological cycle, acts as carbon sinks, pollutant absorption, and reduction of the threat of global warming.

3.4. Biodiversity in India

India is known for its rich heritage of biological diversity, having already documented over 89,000 species of animals and 46,000 species of plants in its 10 bio-geographic regions. Nearly 6,500 native plants are still used prominently in indigenous healthcare systems. Thousands of locally-adapted crop varieties grown traditionally since ancient times, and over 130 native breeds of farm livestock continue to thrive in its diversified farming systems.

India is a land which has four kinds of bio-geographical diversities. They are:

a. Historical

b. Ecological

c. Phytobiological

d. Zoogeographical

Every one of them has its own differences and distinctions. The ten biogeographic regions can be categorized as follows.

1. **The Himalayan Region:** The Himalayan Region covers North-West Himalayas, Western Himalayas, Central Himalayas and the Eastern Himalayas. The altitude gradient of this region has contributed to tremendous biodiversity. The flora and fauna of this region change according to altitude and climatic conditions. The flora of this region consists of tropical rain forests in the Eastern Himalayas and dense sub-tropical and alpine forests in the Central and Western Himalayas. The main trees in the forests of this region are pine, cork tree, sal, castor, etc. The fauna of this region comprises wild bear, sambar, leopard, Sikkim stag, musk deer etc.

2. **The Gangetic Plains:** The Gangetic plains comprise the Upper Ganga Plains and the Lower Ganga Plains, which stretches from Eastern Rajasthan through Uttar Pradesh to Bihar and West Bengal. This region is considered as one of India's most fertile regions. The soil of this region is formed by the alluvial deposits of the Ganges and its tributaries. The flora of this region consists of tropical dry deciduous forests and littoral and mangrove forests. The important trees in the forests of this region are acacia, sal, jamun, mango, etc. The fauna of this region comprises black chinkara, stag, rhinoceros, alligator, turtle, etc. There are tamed animals, birds and vegetations are also present.

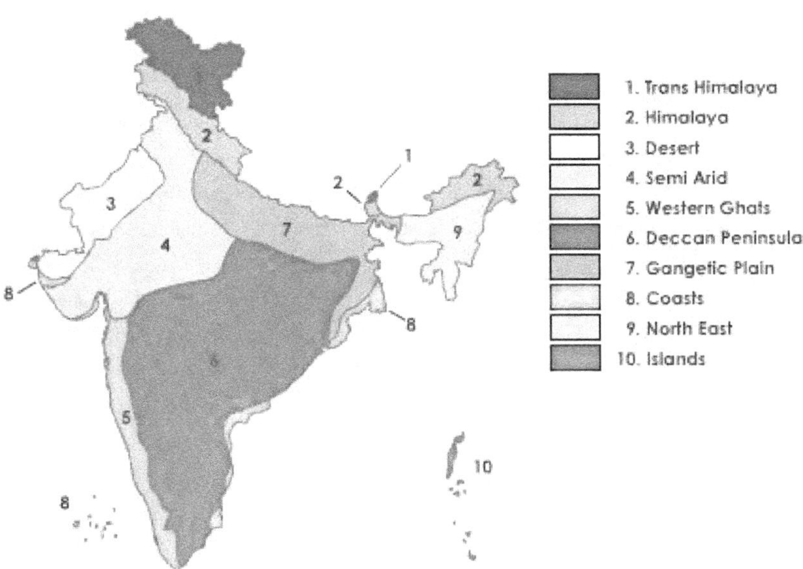

3. **The Desert Zone:** The desert zone consists of the salt desert of Kutch in Gujarat and the sand desert of Thar in Rajasthan. The natural vegetation of this region consists of tropical thorny forests and seasonal salt marshes. The typical trees of this region are acacia, grass, etc. The fauna of this region comprises insects, reptiles, nilgai, wild ass, desert foxes, wolf, chinkara, Indian bustard, desert cat, etc.

4. **The Deccan Plateau:** The Deccan Peninsula is the South and South-Central Plateau, South of the river Tapti. This region includes Central Highlands, Chota Nagpur, Eastern Highlands, Central Plateau and South Deccan. It is a large area of raised land, covering about 43% of India's total land surface. This region is bound by the Satpura range on the North, Western Ghats on the West and the Eastern Ghats on the East. The elevation of the plateau varies from 900m in the west to 300m in the east. There are four major rivers that support the wetlands of this region, which have fertile black and red soils. A large part of this region is covered by tropical forests. Tropical dry deciduous forests occur in the Northern Central and Southern parts of this plateau. The eastern part of the plateau in Andhra Pradesh, Madhya Pradesh and Orissa has moist deciduous forests. The fauna of this region comprises tiger, leopard, sloth bear, wild boar, gaur, samba, wild buffaloes, elephants, etc.

5. **Semi-Arid Zone:** It starts in Rajasthan and extends up to some parts in the state of Punjab and Haryana. In this area, ground water and surface water is much less. Dry xerophytic vegetation is predominant, fauna is also minimum. Very few orchids and bamboo and other plants are seen.

6. **The Coastal Zone:** India has a vast coastal area of about 7516.6 km length. They are the west and east coasts of India. Adjacent to the west coast mangroves and dense forests exist. They are evergreen forests. They are having fine varieties of trees, animals and birds. In the Malabar coastal area Thorium is available. In the east coast, pearl fishing and fishing activities are carried out. The fauna of this zone comprises humpback, inshore dolphin, crocodiles, turtles, tortoises, etc. The highest tiger population is found in the Sundarbans along the east coast adjoining the Bay of Bengal.

7. **The Islands:** The Islands comprise the Andaman and Nicobar Islands in the Bay of Bengal and the Lakshadweep Islands in the Arabian sea. The Andaman and Nicobar Islands are a group of 325 islands. Rainfall is heavy in these islands. The Lakshadweep Islands consists of 30 major islands. The flora of the Andaman and Nicobar Islands comprise jackfruit, coconut, cardamom, clove etc. The fauna of Andaman and Nicobar Islands comprises dolphin, alligator, water snake, etc. The fauna of the Lakshadweep Islands comprises turtles, crabs, sponges, etc.

8. **Trans Himalayan Zone:** This is the northern most area of the country around Himalayas. This zone is not related to mountains but it is the area present surrounding the mountains. This region shows irregular vegetation. It has the richest wild sheep producing quality wool, which is qualitatively and quantitatively superior in the world. Snow leopard is a special animal observable in this zone. Migratory birds like black neck crane are seen here. The great Indian bustard which is an endangered variety is also seen in the grasslands west to this zone.

9. **North East Part of India Zone:** This is the North East Frontier Agency (NEFA) having 7 sister states like Arunachal Pradesh, Meghalaya, Assam, Nagaland, Tripura, Mizoram and Manipur. These states have humid weather and mountainous terrain, having rich culture along with flora and fauna of rare endemic and endangered species of plants and animals in Khasi and Garo hills of Arunachal Pradesh, Meghalaya and Assam. These are the pioneering states that lead in cultivation of tea and coffee plantations.

10. **Western Ghats:** It represents the mountainous western zone of south peninsular India having rich flora and fauna with tropical rain forests extending from Konkan region of Maharashtra up to the western part of Keralagenerally called Malabar coast of Arabian Sea. Wild relatives of cultivated plants like banana, mango, citrus, black pepper are found abundantly in this part.

3.5. Threats to Biodiversity

The loss of biological diversity is a global crisis. The world's tropical forests are disappearing at an alarming rate. Over population, large number of cattle heads, growing demand for land, energy and water supply pose severe problems to Biodiversity. Unplanned developmental works and over exploitation of resources have made the living resources most vulnerable. Over exploitation has not only resulted in shortage of various materials but also left our biodiversity exposed to various ecological threats. The world's biodiversity is under threat from various dangers, the majority of which have been caused by human.

- **Habitat Loss and Fragmentation** is considered by conservation biologists to be the primary cause of biodiversity loss. Clearance of native vegetation for agriculture, housing, timber and industry, as well as draining wetlands and flooding valleys to form reservoirs, destroys these habitats and all the organisms in them. In addition, this destruction can cause remaining habitats to become fragmented and too small for some organisms to persist, or fragments may be too far apart for other organisms to move between.

- **Invasive Alien Species** are the second greatest threat to biodiversity worldwide. Whether introduced on purpose or accidentally, non-native species can cause severe problems in the ecosystems they invade, causing huge changes in ecosystem functioning and the extinction of many species. Virtually all ecosystems worldwide have suffered invasion by the main taxonomic groups. This problem will probably get worse during the next century driven by climate change, and an increase in global trade and tourism as well as the risks to human health, massive economic costs to agriculture, forestry, fisheries and other human activities.

- **Poaching of Animals** is the illegal exploitation of wild species. Commercially, wild animals are hunted for skin, tusks, antlers and fur for meat, pharmaceuticals, perfumes, cosmetics and decoration purposes. Poaching of animals takes place for the following reasons.

  - **Hunting for Money:** Skins of tiger, tusk of elephant, horns of rhinoceros are in great demand and attract many professional hunters as they bring quick and easy money.

  - **Hunting for Pleasure:** The people go to hunting with improved modern aids like jeep, binoculars and flashlight and a number of animals are shot for pleasure.

  - **Bad Hunting Method:** Tribes use poison as a method for hunting which result in the death of many animals.

- **Pollution** is currently poisoning all forms of life, both on land and in water and also contributes to climate change. Any chemical in the wrong place or at the wrong concentration can be considered as pollutant. These chemicals can directly affect biodiversity or lead to chemical imbalance in the environment that ultimately kills individuals, species and habitats. The impact of coastal pollution is also important. Coral reefs are being threatened by pollution. Noise pollution is also the cause of wild life extinction.

- **Deforestation** is one of the main causes for the loss of biodiversity. Deforestation mainly result from population settlement, shifting, cultivation, demand for fuel, wood etc.

- **Climate change** brought about by emissions of greenhouse gases when fossil fuels are burnt, is making life uncomfortably hot for some species and uncomfortably cold for others. This can lead to a change in the abundance and distribution of individual species around the globe and will affect the crops, cause a rise in sea levels and problems to many coastal ecosystems. In addition, the climate is becoming more unpredictable and extreme devastating events are becoming more frequent.

- **Over exploitation** by humans causes massive destruction to natural ecosystems. Exploitation of biodiversity occurs for food (e.g. fish), construction (e.g. trees), industrial products (e.g. animal blubber, skins), the pet trade (e.g. reptiles, fish, orchids), fashion (e.g. fur, ivory) and traditional medicines (e.g. rhino horn). Selective removal of an individual species can unbalance ecosystems and all other organisms within them. In addition, the physical removal of one species often harms other (e.g. fishing by-catches)

- **Man-Wildlife Conflict:** Instances of man animal conflicts keep on increasing in our country. Several instances of killing of elephants in the border regions of Kote-Chamarajanagar belt in Mysore have been reported. The man-elephant conflict in this region is because of the massive damage done by the elephants to the farmer's cotton and sugarcane crops. The agonized villagers electrocute the elephants and sometimes hide explosives in the sugarcane fields, which explode as the elephants intrude into their fields.

- **Human population** is growing at an exponential rate resulting in the above mentioned problems. There are more than 7 billion people in the world although natural disasters, disease and famines cause massive human mortality. Human population tripled in the twentieth century and although growth is slowing, it will take until the twenty-third century to level out at around 11 billion.

3.6. Endangered and Endemic Species of India

3.6.1. *Endangered Species of India*

A plant, animal or microorganism that is in immediate risk of biological extinction is called endangered species or threatened species. In India, 450 plant species have been identified as endangered species. 100 mammals and 150 birds are estimated to be endangered. India's biodiversity is threatened primarily due to,

1. Habitat destruction

2. Degradation

3. Over exploitation of resources

The RED-data book contains a list of endangered species of plants and animals. It contains a list of species of that are endangered but might become extinct in the near future if not protected.

Some of the rarest animals found in India are:

1. Asiatic cheetah

2. Asiatic Lion

3. Asiatic Wild Ass

4. Bengal Fox

5. Gaur

6. Indian Elephant

7. Indian Rhinoceros

8. Marbled Cat

9. Markhor

Extinct species is one which is no longer found in the world. Endangered or threatened species is one whose number has been reduced to a critical number. Unless it is protected and conserved, it is in immediate danger of extinction. Vulnerable species is one whose population is facing continuous decline due to habitat destruction or over exploitation. However, it is still abundant.

Rare species is localized within a restricted area or is thinly scattered over an extensive area. Such species are not endangered or vulnerable. A few endangered species in the world are listed below:

1. West Virginia Spring Salamander (U.S.A)

2. Giant Panda (China)

3. Golden Lion Tamarin (Brazil)

4. Siberian Tiger (Siberia)

5. Mountain Gorilla (Africa)

6. Pine Barrens Tree Frog (Male)

7. Arabian Oryx (Middle East)

8. African Elephant (Africa)

Other important endangered species are:

1. Tortoise, Green sea Turtle, Gharial, Python (Reptiles).

2. Peacock, Siberian White Crane, Pelican, Indian Bustard (Birds).

3. Hoolock gibbon, Lion-tailed Macaque, Capped monkey, Golden monkey (Primates).

4. Rauvolfiaserpentina (medicinal plant), Sandal wood tree, etc.

Factors Affecting Endangered Species

1. Human beings dispose waste indiscriminately into nature thereby polluting the air, land and water. These pollutants enter the food chain and accumulate in living creatures resulting in death.

2. Over-exploitation of natural resources and poaching of wild animals also lead to their extinction.

3. Climate change brought about by accumulation of green house gases in the atmosphere. Climate change threatens organisms and ecosystems and they cannot adjust to the changing environmental conditions leading to their death and extinction.

3.6.2.   Endemic Species of India

Endemics are species that are found in a locality/area and nowhere else in the world. They have a value due to their uniqueness. Area of endemism containing several endemic species, genera or even families has generally been isolated for a long time. Out of about 47,000 species of plants in our country 7,000 are endemic. Indian subcontinent has about 62% endemic flora, restricted mainly to Himalayas, Khasi Hills and Western Ghats. A large number, out of total of 81,000 species of animals in our country is endemic. The Western Ghats are particularly rich in amphibians (frogs, toads etc.) and reptiles (lizards, crocodiles, etc.). About 62% amphibians and 50% lizards are endemic to Western Ghats. Almost 60% the endemic species in India are found in Himalayas and the Western Ghats. Endemic species are mainly concentrated in:

1. North-East India

2. North-West Himalayas

3. Western Ghats and

4. Andaman & Nicobar Islands.

Examples of endemic Flora species are:

1. Sapria Himalayana

2. Ovaria lurida

3. Nepenthiskhasiana etc

Significant endemic fauna in the Western Ghats are:

1. Lion tailed macaque
2. Nilgiri langur
3. Brown palm civet and
4. Nilgiri Tahr

Factors Affecting Endemic Species

1. Habitat loss and fragmentation due to draining and filling of inland wetlands.
2. Pollution plays an important role. Frog eggs, tadpoles and adults are extremely sensitive to pollutants especially pesticides.
3. Over-hunting.
4. Population can be adversely affected by introduction of non-active predators and competitors. Disease producing organisms are also adversary in reducing population of endemic species.

3.7. Conservation of Biodiversity

Biodiversity conservation means, the methods of protecting the diversity of plants and animals that are affected by human activities. Entire ecosystems are preserved by the conservation of a specific species since all species are interdependent. It is essential to save and maintain species and ecosystems for the survival of human beings.

The act or process of conserving i.e. protection, preservation, management, or restoration of wildlife and of natural resources such as forests, soil and water is called conservation of biodiversity.

Conservation of our natural resources has the following three specific objectives:

1. To maintain essential ecological processes and life-supporting systems.
2. To preserve the diversity of species or the range of genetic material found in the organism on the planet.
3. To ensure sustainable utilization of species and ecosystems which support millions of rural communities as well as the major industries all over the world.

Efforts have been made to conserve biodiversity by two basic approaches, such as,

1. In-situ conservation
2. Ex-situ conservation

i) *In-situ Conservation*

Conservation of a species in its natural ecosystem or in a man-made ecosystem is called in-situ conservation. In-situ conservation requires the identification and protection of natural areas that have high biodiversity. This includes conservation and establishment of natural parks and sanctuaries. Large animals like tigers and elephants need large reserves. So, it is necessary to preserve large areas of undeveloped land in which ecosystems and biodiversity can be alive and active.

India has a long history of in-situ conservation of fauna through protected areas. Protected areas include national parks, sanctuaries and biosphere reserves. With the setting up of the enactment of the Wildlife (protection) Act of 1972, the protected areas network has been strengthened. There are 73 national parks and 413 sanctuaries covering about 4% of the total geographic area of the country. "Project Tiger", launched in 1973, succeeded in increasing the tiger population in India. Today in India there are more than 166 national parks and over 500 sanctuaries.

ii) *Ex-situ Conservation*

Conservation of species particularly of endangered species, away from natural habitat under human supervision is called ex-situ conservation. It refers to conservation of species in suitable locations outside their natural habitat. In ex-situ conservation, biodiversity is preserved by artificial means. This includes the storage of seeds in banks, breeding of captive animal species in zoos and setting up of botanical gardens, aquariums and research institutes. When the population of a species is so low that its survival in the wild may not be possible, then steps towards conservation have to be taken.

There are about 100 seed banks in the world and they hold more than four million species of seeds. The seeds are considered to be safe as they are maintained at low temperatures and low humidity levels. There are many problems with seed banks. Sudden fire or power failure can damage seeds permanently. The meristem tips (embryonic tissue that divides fast), buds and stem tips are kept at low temperatures (-3^0 Celsius to 12^0 Celsius) for slow growth and long storage. Field gene banks are the places where collections of growing plants have been assembled to maintain the widest practicable range of biodiversity.

In order to complement the efforts made for in situ conservation, attention has also been given to ex-situ conservation. There are 122 botanical gardens and 33 university – level biological gardens. There are also 205 areas for ex situ wildlife preservation, including 107 zoos, 49 deer parks, thirteen safari parks, six snake parks, 24 nature / education / breeding centres and six aquaria. Some of the major zoos have made significant achievements in the captive breeding of endangered species. To improve the management of zoos, the Government of India has set up the Central Zoo Authority.

The collection and preservation of genetic resources are done through the National Bureau of Plant Genetic Resources in Delhi (for the wild relatives of crop plants), the National Bureau of Animal Genetic Resources in Karnal (for domestic animals) and the National Bureau of Fish Genetic Resources in Allahabad (for economically valuable fish).

The conservation of biological diversity is more than an aesthetic or moral issue; it is integral to our health and economy. The diversity of life is the foundation upon which sustainable development depends. Biodiversity loss can in many cases be attributed to human exploitation of land and resources; however, human intervention can also alter this trend. The United Nations Convention on Biological Diversity is a start towards a responsible course of action. The management of biodiversity is a complex matter that needs the involvement of many different partners ranging from governmental organizations to private companies, NGO's and volunteers. This aside, national and international commitment, legislation and enforcement offer an essential framework for promoting and maintaining biodiversity.

CHAPTER IV

ENVIRONMENTAL POLLUTION

Pollution is the introduction of contaminants into the natural environment that causes adverse changes. The pollutants are waste materials created by the human beings that pollute the natural resources like air, water, soil etc. The chemical nature, concentration and long persistence of the pollutants continually disturb the ecosystem for years. The pollutants can be poisonous gases, pesticides, herbicides, fungicides, organic compounds, radioactive materials, noise etc.

Any type of pollution in our natural surroundings and ecosystem causes insecurity, health disorders and discomfort in normal living. It disorganizes the natural system and thus disturbs the nature's balance.

Following are the different kinds of environmental pollution: Air pollution, Water pollution, Soil pollution, Marine pollution, Noise pollution and Thermal pollution.

4.1. Air Pollution

Air pollution may be defined as the presence of one or more contaminants like dust, mist, smoke and poisonous chemicals in the atmosphere that are harmful to human beings, plants and animals.

4.1.1. *Sources of Air Pollution*

Sources of air pollution are natural sources and artificial sources. Natural sources of pollution are those that are caused due to natural phenomena. Ex: Volcanic eruptions, Forest fires, Biological decay, Pollen grains, Radioactive materials. Artificial sources are those which are created by man. Ex: Thermal power plants, vehicular emissions, fossil fuel burning, agricultural activities etc.

4.1.2. *Causes of Air Pollution*

a. **Burning of Fossil Fuels:** Sulphur dioxide emitted from the combustion of fossil fuels like coal, petroleum and other factory combustibles is one of the major causes of air pollution. Pollutants from vehicles including trucks, jeeps, cars, trains, aeroplanes cause immense amount of pollution. Carbon Monoxide released from vehicles due to improper or incomplete combustion is another major pollutant.

b. **Agricultural Activities:** Ammonia is a very common by-product from agriculture related activities and is one of the most hazardous gases in the atmosphere.

c. **Exhaust from Factories and Industries:** Manufacturing industries release large amount of carbon monoxide, hydrocarbons, organic compounds and chemicals into the atmosphere there by depleting the quality of air. Petroleum refineries release hydrocarbons and various other chemicals that pollute the air.

d. **Mining Operations:** Mining is a process wherein minerals below the earth are extracted using huge equipments. During the process, dust and chemicals are released into the air causing massive air pollution.

e. **Indoor Air Pollution:** Household electronic gadgets such as air conditioners, refrigerators and household cleaning products, paints release toxic chemicals into the atmosphere and cause air pollution.

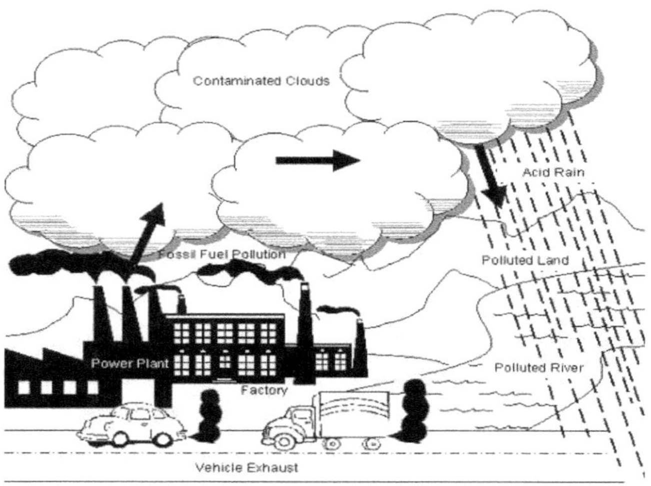

4.1.3. Effects of Air Pollution

a. **Respiratory and Heart Problems:** They are known to create several respiratory and heart problems along with cancer and other severe threats to the human beings. Children in the areas exposed to air pollutants are said to commonly suffer from pneumonia and asthma. Exhaust from factories and industries are one of the reasons for the deteriorating health conditions of workers and nearby residents.

b. **Global Warming:** Air pollutants are major contributors to climate change and Global warming.

c. **Acid Rain:** Harmful gases like Nitrogen oxide and Sulphur oxide are released into the atmosphere during the burning of fossil fuels. When it rains, the water droplets combines with these air pollutants becomes acidic and then falls on the ground in the form of acid rain. Acid rain can cause great damage to human beings, animals and crops.

d. **Effect on Wildlife:** Toxic chemicals present in the air can force wildlife species to move to a new place and change their habitat. The toxic pollutants deposited over the surface of the water can also affect the sea animals.

e. **Depletion of Ozone Layer:** Ozone layer exist in the earth's stratosphere is responsible for protecting life on earth from harmful ultraviolet (UV) rays. Earth's ozone layer is depleting due to the presence of chlorofluorocarbons, hydro chlorofluorocarbons in the atmosphere. As ozone layer becomes thin, it allows harmful rays to enter into earth's surface and can cause skin and eye related problems.

4.1.4. *Control Measures for Air Pollution*

a. **Using Public Mode of Transportation:** People should be encouraged to use more and more public modes of transportation and eco-friendly vehicles to reduce pollution and also to save energy.

b. **Conserving Energy:** Large amount of fossil fuels are burnt to produce electricity. Optimum utilization of electronic gadgets and using energy efficient devices will help to reduce pollution by consuming less energy.

c. **Understanding the Concept of "Reduce, Reuse and Recycle":** Instead of throwing away items that are considered as no use, can be reused or recycled for some other purpose.

d. **Emphasis on Clean Energy Resources:** Use of clean energy sources like solar, wind and geothermal energy (green energy) should be encouraged through grants and subsidies.

4.2. Water Pollution

Water pollution is the pollution or contamination of natural water bodies like lakes, rivers, streams, oceans and groundwater due to inflow of pollutants into water systems. This brings about changes in the chemical, physical and biological properties of water that can cause harmful consequences on living things and the environment.

4.2.1. Causes of Water Pollution

Industrial activities cause heavy water pollution. Waste from factories is let off into fresh water then into rivers. This contaminates water with pollutants like lead, mercury and petrochemicals. The significant contributors of wastewater are paper mills, steel plants, textile dyeing and sugar industries. Water pollution is due to:

a. Sewage let off from domestic households, factories, commercial buildings containing chemicals and pharmaceuticals which are untreated and disposed into the sea cause greater problems.

b. Solid waste dumping in water bodies.

c. Oil spills from tankers and ships because water pollution as oil does not dissolve in water and forms a thick layer on the water surface.

d. Detergents, food processing waste, insecticides, petrochemicals, debris from logging operations, drugs, chemical wastes, fertilizers and heavy metals are other causes of water pollution.

4.2.2. *Effects of Water Pollution*

Water pollution extensively affects health in human beings and aquatic ecosystems.

a. Groundwater contamination causes reproductive and fertility disorders in wildlife ecosystems and makes it unpalatable for human consumption.

b. Sewage, fertilizer and agricultural run-off have organic substances which lead to algal bloom causing oxygen depletion. The lower oxygen levels affect the natural ecological balance of river and lake ecosystems. Both the Biological Oxygen Demand (BOD) and Chemical Oxygen Demand (COD) tests are a measure of the relative oxygen-depletion effect of a waste contaminant. Both have been widely adopted as a measure of pollution effect. The BOD test measures the oxygen demand of biodegradable pollutants whereas the COD test measures the oxygen demand of biodegradable pollutants plus the oxygen demand of non-biodegradable oxidizable pollutants.

c. Consumption of contaminated water causes skin diseases, cancer, reproductive problems and stomach ailment in humans.

d. Industrial effluents and agricultural pesticides that accumulate in aquatic environment cause harm to aquatic animals and lead to biomagnifications. Heavy metals like mercury, lead are poisonous to human beings as these chemicals affect the development of nervous system.

e. Littering of plastic by humans affect aquatic animals.

   E.g: Minamata disease was first discovered in Japan, in 1956. It was caused by the release of methylmercury in the industrial wastewater from the chemical factory into the sea. This highly toxic chemical bioaccumulated in fish which, when eaten by people, resulted in mercury poisoning, which led to death of cat, dog, pig, and human beings.

4.2.3. Control Measures for Water Pollution

a. Water pollution can be controlled to a large extent on the principle, "the solution to pollution is dilution."

b. The sewage pollutants are subjected to chemical treatment to change them into non-toxic substances or make them less toxic.

c. Water pollution due to organic insecticides can be reduced by the use of very specific and less stable chemicals in the manufacture of insecticides.

d. Domestic and industrial wastes must be stored in large but shallow ponds for some days. Due to the sun-light and the organic nutrients present in the waste there will be mass scale growth of those bacteria which will digest the harmful waste matter.

e. Polluted water can be reclaimed by proper sewage treatment plants and the same water can be reused in factories and even for irrigation. Treated water being rich in phosphorus, potassium and nitrogen can be a good fertilizer.

f. Suitable strict legislation should be enacted to make it obligatory for the industries to treat the waste water before being discharged into rivers or seas.

4.3. Soil Pollution

Soil pollution is defined as the contamination of soil in a particular region. Soil pollution is the result of penetration of harmful pesticides and insecticides that deteriorate the soil quality making it unfit for later use. This is usually caused by industrial activity, agricultural chemicals and improper disposal of waste. The most common pollutants are petroleum, hydrocarbons, solvents, pesticides, lead and other heavy metals.

4.3.1. Causes for Soil Pollution

Soil Pollution is a result of many activities by mankind which contaminate the soil.

Some of the causes of soil pollution are as follows:

a. Industrial effluents and chemicals.

b. Improper or ineffective soil management system.

c. Unfavorable irrigation practices.

d. Improper management and maintenance of septic system.

e. Sanitary waste leakage.

f. Acid rain.

g. Leakages of fuel from automobiles are washed off due to rain and are deposited in the nearby soil.

h.   Unhealthy waste management techniques release sewage into dumping grounds and nearby water bodies.

i.   Use of pesticides in agriculture retains chemicals in the environment for a long time. These chemicals also affect beneficial organisms (non-target organisms) like earthworm in the soil and lead to poor soil quality.

j.   Absence of proper garbage disposal system leads to scattered garbage in the soil. These contaminants can block passage of water into the soil and affects its water holding capacity.

k.   Unscientific disposal of nuclear waste contaminate soil and can cause mutations in organisms.

l.   Night soil contamination due to improper sanitary system in villages can cause harmful diseases.

4.3.2. *Effects of Soil Pollution*

Soil pollution causes enormous disturbances in the ecological balance and health of living organisms at an alarming rate. Some of the effects of soil pollution are:

a.   Reduced soil fertility and hence decrease in the yield.

b.   Loss of natural nutrients in the soil.

c.   Reduced nitrogen fixation.

d.   Increased soil erosion.

e.   Imbalance in the flora and fauna of the soil.

f.   Increase in soil salinity, makes it unfit for cultivation.

g.   Creation of toxic dust.

h.   Alteration in soil structure that lead to death of organisms in it.

4.3.3. *Control Measures for Soil Pollution*

The soil pollution can be controlled by:

a.   Recycling paper, plastics and other materials.

b.   Ban on use of plastic bags.

c.   Reusing materials.

d.   Avoiding deforestation and promoting afforestation.

e.   Suitable and safe disposal of wastes including nuclear waste.

f.   Chemical fertilizers and pesticides should be replaced by organic fertilizers and pesticides.

g. Encouraging social and agro forestry programmes.

h. Undertaking pollution awareness programmes.

4.4. Noise Pollution

Noise is a sound which is unwanted at the wrong place and at the wrong time. The unit of measurement of intensity of sound is decibel (dB). Human beings are very sensitive to a wide range of intensity from 0 to 180dB. 0dB is said to be the threshold of hearing while 140dB marks the threshold of pain.

Noise pollution is defined as the disturbing or excessive noise that may cause harm to the activities of human beings and animals.

4.4.1. *Causes for Noise Pollution*

a. **Household Sources:** Electrical gadgets like grinder, mixer, vacuum cleaner, washing machine, dryer, cooler and air conditioners produce more noise. Other domestic sources that cause noise pollution can be speakers of sound systems, televisions, iPods and earphones.

b. **Social Events:** Places of worship and social events like discos, gigs, parties create a lot of noise. In Market areas, people sell with loud voices and speakers to lure customers causing noise.

c. **Commercial and Industrial Activities:** Industries like printing press, manufacturing industries, construction sites contribute a lot to noise pollution. Equipments like lawn mowers and tractors also cause noise.

d. **Transportation:** Aeroplanes, jets, vehicles on road, trains constantly create noise pollution.

4.4.2. *Effects of Noise Pollution*

The effects caused by noise pollution are as follows:

Effects on Human Beings

a. Long exposure to loud noise may result in hearing impairment which may become permanent.

b. It causes anxiety and stress reaction and in extreme cases fear and an increase of heart rate, increase in the cholesterol level and blood pressure.

c. It can cause constriction of blood vessels.

    d. Nervousness, headache, irritability, fatigue and decrease in work efficiency can be caused due to noise pollution.

    e. It also affects the development of embryo in mother's womb.

Effects on Wildlife

    a. Noise pollution can have harmful effects on wildlife. It can lead to changes in the delicate balance in predator and prey detection.

    b. It also interferes with the sounds of communication and in the relation to reproduction and navigation.

    c. Over exposure to noise can lead to temporary or permanent loss of hearing.

4.4.3. Control Measures for Noise Pollution

    a. Plantation of plant bushes and trees around the sound generating sources as trees and plants block sound passage.

    b. Sound proof doors and windows can be installed.

    c. Strict laws are essential to regulate the usage of loud speakers in public areas.

    d. Construction of factories away from residential areas.

    e. Using technology like white noise which converts unbearable noise into pleasant sound.

    f. Machinery has to be lubricated in order to minimize noise generation and the machine quality can be optimized to reduce production of sound.

    g. Noise from music systems and television sets can be kept at moderate level.

    h. Ear protection devices are to be worn while working in high noise levels.

    i. Vehicles and machinery have to be maintained properly and checked from time to time.

    j. Sound insulation can be installed at the roof top, in order to reduce the aircraft noise.

4.5. Thermal Pollution

Thermal pollution is defined as the addition of excess undesirable heat to air, land and water by human activities that are harmful to man, animal or aquatic life.

4.5.1. *Sources of Thermal Pollution*

The main contributors to thermal pollution are:

a. Nuclear power plants

b. Coal fired plants

c. Industrial effluents

d. Domestic sewage

e. Hydro-electric power

    a. **Nuclear Power Plants:** Nuclear power plants including drainage from hospitals, research institutions, nuclear reactors discharge a lot of heat. The operations of power reactors and nuclear fuel processing units are the major contributors of heat in the aquatic environment. Heated effluents from power plants are discharged at 10°C higher than the receiving water and it affects the aquatic flora and fauna.

    b. **Coal-fired Power Plants:** Coal fired power plants constitute a major source of thermal pollution. The condenser coils in such plants are cooled with water from nearby lakes or rivers. The resulting heated water is discharged into streams thereby raising the water temperature by 15°C. Heated effluent decreases the dissolved content of water, resulting in death of fish and other aquatic organisms.

    c. **Industrial Effluents:** Industries like textile, paper, pulp and sugar manufacturing units release huge amount of cooling water along with effluents into nearby natural water bodies and thus water is polluted by sudden and heavy organic loads, resulting in severe drop in the levels of dissolved oxygen.

    d. **Hydro-electric Power:** Generation of hydroelectric power sometimes leads to negative thermal loading in water systems.

    e. **Natural Causes:** Natural causes like volcanoes and geothermal activities under the oceans and seas can trigger warm lava to raise the temperature of water bodies. Lightening can also introduce massive amount of heat into the oceans. This means that the overall temperature of the water source will rise, having significant impact on the environment.

4.5.2. *Effects of Thermal Pollution*

The effects of thermal pollution on ecosystem are as follows:

a. **Decrease in DO (Dissolved Oxygen) Levels:** The warm temperature reduces the levels of DO (Dissolved Oxygen) in water. The warm water holds relatively less oxygen than cold water. The decrease in DO can create suffocation for plants and animals such as fish, amphibians and copepods, which may give rise to anaerobic conditions. Warmer water allows algae to flourish on surface of water and over the long term, growing algae can decrease oxygen levels in the water.

b. **Increase in Toxins:** With the constant flow of high temperature discharge from industries, there is a huge increase in toxins that are being regurgitated into the natural body of water. These toxins may contain chemicals or radiation that may have harsh impact on the local ecology and make them susceptible to various diseases.

c. **Loss of Biodiversity:** A dent in the biological activity in the water may cause significant loss of biodiversity. Changes in the environment may cause certain species of organisms to shift their base to some other place while there could be significant number of species that may shift in because of warmer water. Organisms that can adapt easily may have an advantage over organisms that are not used to the warmer temperatures.

d. **Ecological Impact:** A sudden thermal shock can result in mass killings of fish, insects, plants or amphibians. Many aquatic species are sensitive to small temperature changes such as one degree Celsius that can cause significant changes in metabolism and other adverse cellular biological effects.

e. **Interference in Reproduction:** A significant halt in the reproduction of marine wildlife can happen due to the increase in temperature as reproduction can happen within certain range of temperature. Excessive temperature can cause the release of immature eggs or can prevent normal development of certain eggs.

f. **Increases Metabolic Rate:** Thermal pollution increases the metabolic rate of organisms that increases the enzyme activity that causes organisms to consume more food. It disrupts the stability of food chain and alters the balance of species composition.

g. **Migration:** The warm water can also cause particular species of organisms to migrate to suitable environment that would cater to its requirements for survival. This can result in loss for those species that depend on them for their daily food as their food chain is interrupted.

4.5.3. *Control Measures for Thermal Pollution*

The following methods can be adapted to control high temperature caused by thermal discharges:

a. **Cooling Towers:** Cooling towers transfer heat from hot water to the atmosphere by evaporation.

b. **Cooling Ponds:** Cooling ponds are the best way to cool thermal discharges.

c. **Spray Ponds:** The water coming out from condensers is allowed to pass into the ponds through sprayers. Here water is sprayed through nozzles as fine droplets. Heat from the fine droplets gets dissipated into the atmosphere.

d. **Artificial Lakes:** Artificial lakes are manmade water bodies. The heated effluents can be discharged into the lake at one end and water for cooling purposes may be withdrawn from the other end. The heat is eventually dissipated through evaporation.

4.6. Marine Pollution

Oceans are the largest water bodies on the Earth. Over the last few decades, human activities have severely affected the marine life in oceans. Marine pollution also known as Ocean pollution is spreading of harmful substances such as oil, plastic, industrial and agricultural waste and chemical particles into the ocean. Since oceans provide home to wide variety of marine animals and plants, it is responsibility of every citizen to keep these oceans clean so that marine species can thrive for a long period of time.

4.6.1. *Sources of Marine Pollution*

a. Waste and sewage from residences and hotels in coastal towns are directly discharged into sea.

b. Pollution from the atmosphere is a huge source of ocean pollution. This occurs when objects like dust, man-made objects such as debris and trash that are far inland are blown by the wind over long distances and end up in the ocean. Most debris, especially plastic debris, cannot decompose and remains suspended in the ocean current for years.

c.   Pesticides and fertilizers from agriculture which are washed off by rain, enter water courses and finally to sea.

d.   Petroleum and oil washed off from roads normally enter sewage system and finally into sea.

e.   Ship accidents and accidental oil spillage in sea affect the marine environment.

f.   Off shore oil exploration pollute the sea water to a large extent.

g.   Residual oil goes into sea while cleaning the hulls and dry docking servicing when cargo compartments are emptied.

h.   Volcanic eruptions in the sea.

i.   Ocean mining in the deep sea.

4.6.2. *Effects of Marine Pollution*

a.   Animals mistake plastic as food and when eaten, it's slowly kills them. Animals are most often the victims of plastic debris that include turtles, dolphins, fish, sharks, crabs, sea birds, and crocodiles.

b.   Apart from causing Eutrophication, a large amount of organic wastes can also result in the development of 'red tides'. These are phytoplankton blooms because of which the whole area is discoloured.

c.   Commercially important marine species are also killed due to clogging of gills and other structures.

d.   When oil is spilled on the sea, it spreads over the surface of the water to form a thin film called as oil slick and damages marine life to a large extent. Commercial damage to fish by tainting which gives unpleasant flavour to fish and sea food that reduces the market values of sea food and also causes the death of birds.

e.   Organic waste addition results in end products such as hydrogen sulphide, ammonia and methane which are toxic to many organisms. Most life disappears except for anaerobic microorganisms and renders the water foul smelling.

f.   The coral reefs are the productive ecosystem which offers many benefits to people. These coral reefs are threatened by the sediments from deforestation carried by the runoffs and the agricultural and industrial chemicals reaching through river discharges.

g.   Industrial and agricultural wastes include various poisonous chemicals that are considered hazardous for marine life as it affects the reproductive system of sea animals.

h. Chemicals used in industries and agriculture get washed into the rivers and from there are carried into the oceans. These chemicals do not get dissolved and sink at the bottom of the ocean. Small animals ingest these chemicals and are later eaten by large animals, which then affect the whole food chain.

i. Animals from affected food chain are then eaten by humans which affects their health as toxins from these contaminated animals gets deposited in the tissues of people and can lead to cancer, birth defects or long term health problems.

4.6.3. *Control Measures for Marine Pollution*

a. Introduction of sewage treatment plants to reduce BOD of final product before discharging into sea.

b. Cleaning oil from surface water and contaminated beaches can be accelerated through the use of chemical dispersants which can be sprayed on the oil.

c. Skimming off the oil on water surface with a suction device.

d. Spreading a high-density powder over the oil spill, so that oil can be sunk to the bottom.

4.7. Role of an Individual in the Prevention of Pollution

Pollution Prevention is the use of processes, practices, materials, products, substances or energy that avoids or minimizes the creation of pollutants and waste and reduces the overall risk to human health or the environment. People can prevent pollution by changing how they do things and by changing the materials they use. Pollution prevention is by "reducing, reusing and recycling" through individual action, government policy and industrial and business actions.

a. *Decreasing Waste*

1. The concept of pollution prevention is based on the idea of eliminating the creation of pollutants. In some cases reducing waste is a crucial step in preventing pollution.

2. Choosing products with recyclable packaging or less packaging reduces waste.

3. Avoiding products that cannot be recycled locally.

4. Food wastes can be composed to reduce landfill space and pollution.

b. *Reducing Toxins*

1. Mercury, a bio accumulating toxin, can be found in mercury electrical switches, fluorescent bulbs, thermometers, thermostats and older batteries. To prevent mercury pollution, mercury-free products, like digital thermometers can be used.

2. Degreasers, pesticide formulations, cleaning products and paints often contain ethoxylates and nonylphenol which pollute groundwater. Choosing products free of these chemicals prevent pollution.

c. *Using More Efficient Transportation*

1. Selecting a fuel-efficient vehicle is an essential step for pollution prevention.
2. Maintaining the vehicle is also essential to prevent pollution.
3. Carpooling, public transportation, walking and cycling are ways to reduce pollutants and save money.

d. *Reducing Energy Consumption*

1. Unplugging electronic items when they are not in use and turning off unnecessary lights to reduce energy consumption.
2. Purchasing energy-efficient appliances, electronics devices reduces not only energy consumption but also electricity bills (eg. CFL bulbs).
3. Looking for opportunities to support alternative energy sources. Local electric companies may offer wind or solar power alternatives.

4.8. Solid Waste Management

Solid waste management is a polite term for garbage management. As long as humans have been living in settled communities, solid waste or garbage has been an issue and modern societies generate far more solid waste than early humans ever did. Daily life in industrialized nations can generate several pounds of solid waste per consumer not only directly in the home, but indirectly in factories that manufacture goods.

4.8.1. *Types and Sources of Solid Waste*

Basically solid waste can be classified into different types depending on their source:

Source	Typical Waste Generators	Types of Solid Wastes
Residential	Single and multifamily dwellings	Food waste, paper, plastics, card board, glass, metals, batteries, household hazardous materials.
Industrial	Light and heavy manufacturing fabrication, construction sites, power and chemical plants	Housekeeping waste, packaging, hazardous wastes, ashes.
Commercial and Institutional	Stores, hotels, restaurants, markets, office buildings Schools, hospitals, prisons, government concerns.	Paper, plastics, card board, glass, metals.
Construction and demolition	Road repair, renovation sites, demolition of buildings.	Wood, steel, concrete, dirt etc.
Municipal services	Street cleaning, waste water treatments.	Street sweepings, tree trimming, sludge.
Agriculture	Crops, orchards, dairies, vineyards, farms, feedlots.	Agricultural wastes, hazardous wastes.

4.8.2. Effects of Solid Waste Pollution

Solid wastes heap up on the roads due to improper disposal system. This type of dumping allows biodegradable materials to decompose under uncontrolled and unhygienic conditions. This produces foul smell and breeds various types of insects and infectious organisms besides spoiling the aesthetics of the site.

Industrial solid wastes are sources of toxic metals and hazardous wastes and affect the productivity of soils.

Toxic substances may leach or percolate to contaminate the ground water Various types of wastes like cans, pesticides, cleaning solvents, batteries (zinc, lead or mercury), radioactive materials, plastics and e-waste are mixed up with paper, scraps and other non-toxic materials could be recycled. Burning of some of these materials produces harmful chemicals which have the potential to cause various typesof ailments including cancer.

4.8.3. Solid Waste Management Practices

Solid waste management includes many steps like collection of the waste, its transport, processing, recycling or disposal and monitoring of the waste material and relevant processes/ activities. The system implemented for solid waste management mostly depends on quantity and complexity of the waste materials. There are three main types of waste management methods widely used across the world – Landfill, Incineration and Recycling. Various municipal corporations and waste management companies are involved in these activities.

Landfill: A landfill, also known as a dump site for the disposal of waste materials by burial under the waste management procedures. It is the most common method of organized waste disposal. Landfill for the waste material is associated with many severe problems such as land and groundwater contamination, engagement of land which would have been otherwise useful for the agriculture/other infrastructural activities release of methane which is a potent greenhouse gas.

Incineration: It involves the combustion of organic substances contained in waste materials which further converts the waste into ash, flue gas, and heat. Flue gases involve various pollution gases like oxides of Sulphur, oxides of nitrogen, etc. Some of these gases cause greenhouse effect resulting in climate change and global warming.

Recycling: It includes collection, processing and utilization of waste material. Conversion of waste materials into new products/potentially useful materials reduces the consumption of fresh raw materials. It subsequently results in natural resource conservation.

Varieties of environmental legislation are available in India to treat and manage waste materials. Environmental protection acts encourage and reward organizations / companies for managing and recycling their waste to maintain a clean and hygienic environment.

4.9. Disaster Management

Disaster is a sudden, calamitous event that seriously disrupts the functioning of a community or society that causes human, material and economic or environmental losses. It is a grave misfortune and has severe consequences.

Types of Disasters

A disaster is a natural or man-made hazard resulting in an event of substantial extent causing significant physical damage or destruction, loss of life, or drastic change to the environment. A disaster can be defined as any tragic event stemming from events such as earthquakes, floods, catastrophic accidents, fires, or explosions. It is a phenomenon that can cause damage to life and property and destroy the economic, social and cultural life of people.

Hazards are divided into natural or human-made.

a. **A Natural Disaster** is a natural hazard that affects humans and/or the built environment. Various phenomena like earthquakes, landslides, volcanic eruptions, floods and cyclones are all natural hazards that kill thousands of people and destroy crores of rupees of living environment and property each year. Developing countries suffer more or less chronically by natural disasters. Asia tops the list of casualties due to natural disasters.

b. **Man-made Disasters** are the consequence of technological or human hazards such as stampedes, fires, transport accidents, oil and chemical spills, nuclear radiations and wars. Some examples of worst manmade disasters are:

   - The Bhopal disaster is the world's worst industrial disaster to date.

   - On June 28, 2014, a eleven storied under-construction building at Moulivakkam in the suburb of Chennai, Tamil Nadu collapsed, killing 61 people, mostly construction workers.

   - Fukushima Daiichi nuclear disaster was a nuclear disaster at the Fukushima I Nuclear Power Plant that began on 11 March 2011 and resulted in nuclear meltdown of three of the plant's six nuclear reactors.

4.9.1. *Disaster Management Cycle*

The Disaster management cycle illustrates the ongoing process by which government, business establishment and civil society, plans to reduce the impact of disasters, react during and immediately following a disaster, and take steps to recover after a disaster has occurred. Appropriate actions at all points in the cycle lead to greater preparedness, better warnings, reduced vulnerability or the prevention of disasters during the next iteration of the cycle. The complete disaster management cycle includes the shaping of public policies and plans that either modify the causes of disasters or mitigate their effects on people, property and infrastructure.

The disaster cycle or the disaster life cycle consists of the steps that emergency managers take in planning for and responding to disasters. Each step in the disaster cycle correlates to part of the ongoing cycle that is emergency management. This disaster cycle is used throughout the emergency management community, from the local to the national and international levels.

The first step of the disaster cycle is usually considered to be preparedness. Prior to a disaster's occurrence, emergency manager will plan for various disasters which could strike within the area of responsibility.

The second stage in the disaster cycle is response. Imminently prior to a disaster, warnings are issued and evacuations or sheltering in place occurs and necessary equipment is placed at the ready.

After the immediate response phase of the disaster cycle has been completed, the disaster turns toward recovery, focusing on the longer term response to the disaster. During the recovery phase of the disaster cycle, officials are interested in cleanup and rebuilding. During the recovery phase, lessons learned are collected and shared within the emergency response community.

The mitigation phase of the disaster cycle is almost concurrent with the recovery phase. The goal of the mitigation phase is to prevent the same disaster-caused damages from occurring again.

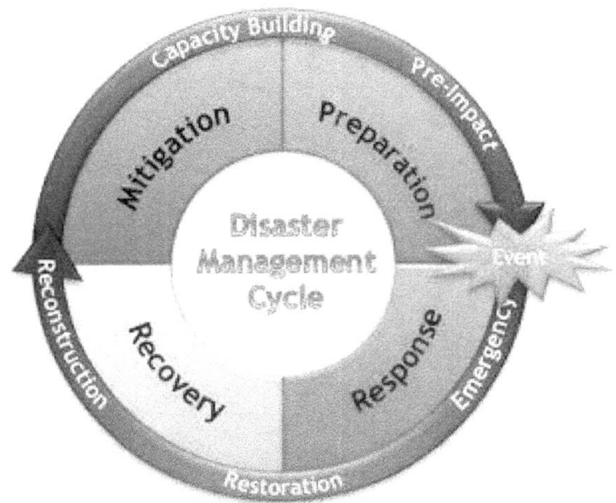

Finally, using the lessons learned from the response, recovery and mitigation phases of the disaster the emergency manager and government officials return to the preparedness phase and revise their plans and their understanding of the material and human resources needs for a particular disaster in their community.

CHAPTER V

SOCIAL ISSUES AND THE ENVIRONMENT

There are several social issues connected with the environment. The important social issues connected with the environment are presented in this chapter.

5.1. Water Conservation

The process of saving water for future utilization is known as water conservation. The principal source of fresh water is rain and is considered to be the purest form. The other sources of water are river, streams, lakes, ponds and also ocean. Since water is the most important component to human survival and all other commercial, agricultural activities, it is essential to conserve the water resources.

5.1.1. Need for Water Conservation

Water Conservation has become mandatory due to the following reasons.

- Though the resources of water are more, the quality and reliability are not high due to change in environmental factors.
- Better lifestyles require more fresh water.
- As the population increases, the requirement of water is also more.
- Annual rainfall is decreasing, due to deforestation.
- Over exploitation of ground water, leads to drought.
- Agricultural and industrial activities require more fresh water.

5.1.2. Strategies for Water Conservation

The following strategies can be adopted for the conservation of water.

a. **Reducing Evaporation Losses:** Evaporation of water in humid region can be reduced by placing horizontal barriers of asphalt below the soil surface, which increases the water availability and crop yield.

b. **Reducing Irrigation Losses:** The water loss during irrigation can be reduced by the following methods.

  - Sprinkler irrigation and drip irrigation conserves water by 30-40%.
  - Growing hybrid crop varieties, which require less water, also conserve water.
  - Irrigation in early morning or later evening reduces evaporation losses.

c. **Re-use of Water**

- Treated waste water can also be used for fertile-irrigation.

- Greywater is gently used water from bathroom sinks, showers, tubs, and washing machines. Grey water from washings, bath-rooms, etc., may be used for washing cars, watering gardens.

d. **Preventing Wastage of Water:** Wastage of water can be prevented by,

- Closing the taps when not in use.

- Repairing leaking pipes.

- Using small capacity taps.

e. **Decreasing Run-off Losses:** Water Run-off on the soil can be reduced by allowing most of the water to infiltrate into the soil. This can be done by using contour cultivation or terrace farming.

f. **Avoiding Discharge of Sewage:** The discharge of sewage into natural water resources should be prevented.

5.1.3. *Methods of Water Conservation*

There are so many methods available for water conservation, of which the following are the important methods.

1. Rain water harvesting
2. Watershed management

1. *Rain Water Harvesting*

Rain water harvesting is a technique of capturing and storing of rainwater for further utilization.

Objectives of Rainwater Harvesting

a. To meet the increasing demands of water.

b. To raise the water table by recharging the ground water.

c. To reduce the ground water contamination from the intrusion of saline water.

d. To reduce the surface run-off loss.

e. To reduce storm water runoff and soil erosion.

f. To increase hydrostatic pressure to stop land subsidence.

g. To minimize water crisis and water conflicts.

The annual average rainfall in India is 1200 mm, that too it is concentrated only in the months of June to November. There is only little vegetation to check the runoff and allow infiltration. India's ground water is not in a very good state. The annual recharge of water is much less than what is consumed. The situation is worst in urban areas where most of the earth's surface has been converted into percolation proof by roads and concrete buildings. So, great effect must be taken to harvest rain water in proper and effective ways.

Concept of Rainwater Harvesting

Rainwater harvesting involves collecting water that falls on the roofs of buildings during rain and conveying it through PVC (or) aluminum pipes to a nearby covered storage unit. The rainwater yield varies with the size and texture of the catchment area. A smoother, cleaner and more impervious roofing material contributes to better water quality and greater quantity.

Advantages of Rainwater Harvesting

    a.   Mitigating the effects of droughts and achieving drought proofing.

    b.   Rise in ground water levels.

    c.   Minimizing the soil erosion and flood hazards.

    d.   Upgrading the social and environmental status.

    e.   Future generation is assured of water.

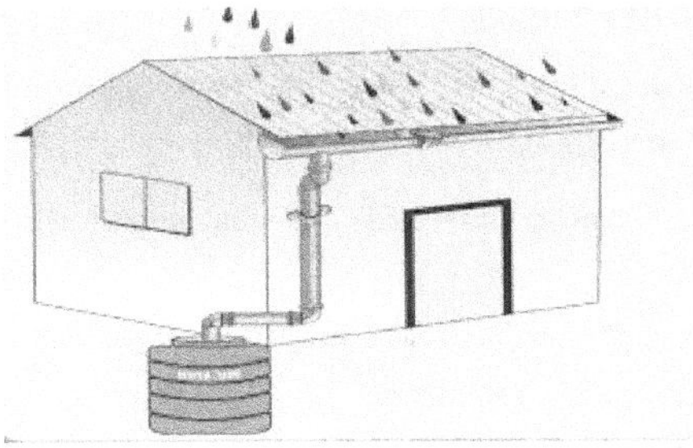

2. Watershed Management

Watershed is defined as the land area from which water drains under the influence of gravity into streams, lakes, reservoirs or other bodies of surface water. The management of rainfall and resultant runoff is called watershed management. It also involves conservation, regeneration and proper use of water. Watershed is not a technology but a concept which integrates construction management and budgeting of rainwater through simple but discrete hydrological units.

Factors Affecting Watershed

1. The watersheds are found to be degraded due to uncontrolled, unplanned and unscientific land use activities.
2. Overgrazing, deforestation, mining, construction activities also affect and degrade various watersheds.
3. Droughty climates also affect the watersheds.

For example our water regimes in Himalayan ranges are threatened resulting in the depletion of water resources due to damage of reservoirs, irrigation systems and misuse of slopes of the mountain. A vast hydroelectric power potential can be harnessed from Himalayan watersheds only if proper control measures are taken.

Need for Watershed Management

1. To minimize the risks of floods, droughts and landslides.
2. To manage the watershed for developmental activities like domestic water supply, irrigation, hydropower generation, etc.,
3. To promote social forestry and horticultural activity on all suitable areas of land.
4. To protect the soil from erosion by runoff
5. To raise the groundwater level.

Watershed Management Techniques

In watershed management, various civil structures were constructed in the catchment area to improve groundwater storage.

1. **Trenches (Pits):** Trenches are made at equal intervals to improve groundwater storage.
2. **Earthen Dam (or) Stone Embankment:** To check the run-off water, earthen dam must be constructed in the catchment area.
3. **Farm Pond:** A farm pond can be built to improve water storage capacity of the catchment area.
4. **Underground Barriers (Dykes):** Underground barriers should be built along the nullahs to raise the water table.

Components of Integrated Watershed Management

1. **Water Harvesting:** Proper storage of water in watershed is done with provisions that the water can be used in dry seasons in low rainfall areas.

2. **Afforestation and Agroforestry:** Afforestation and Agroforestry help to prevent soil erosion and retention of moisture in watershed areas.

3. **Reducing Soil Erosion:** Terracing, bunding, contour cropping, strip cropping, etc., are used to minimize soil erosion and runoff on the slopes of watersheds.

4. **Scientific Mining and Quarrying:** Due to improper mining, the stability of the hills gets disturbed resulting in landslides and rapid soil erosion. Planting soil-binding plants, contour trenching at an interval of 1 meter on overburden dump in the mined area are good for minimizing the destructive effects of mining in watershed areas.

5. **Public Participation:** People's co-operation and participation is essential for watershed management. People must be motivated for protecting freshly planted areas and maintaining a water harvesting structure, implemented by the government.

6. **Minimizing Livestock Population:** Livestock population present in the surrounding villages of the watershed should be reduced.

5.2. Climate Change

Climate is the average weather of an area. It is the general weather conditions, seasonal variations of a region. The average of such conditions over a long period is called climate. The earth's average surface temperature and climate have been changing throughout the world's 4.7 billion-year history. Sometimes it has changed gradually and at other times quickly. However the average temperature has fluctuated by 0.5 - 1°C. India has relatively stable climate for thousands of years based on which we have practiced our agriculture.

Causes of Climate Change

1. Presence of green house gases (carbon dioxide, methane, nitrous oxide, and ozone) in the atmosphere increases the global temperature.

2. Depletion of ozone layer also increases the global temperature.

Effects of Climate Change

1. Even small changes in climatic conditions may disturb agriculture that would lead to migration of animals and human beings.

2. Climate change may upset the hydrological cycle, resulting in floods and droughts in different regions of the world.

3. Global pattern of winds and ocean currents also gets disturbed by climate.

5.3. Global Warming

The greenhouse effect may be defined as, "The progressive warming up of the earth's surface due to blanketing effect of manmade CO_2 in the atmosphere." The increased inputs of CO_2 and other greenhouse gases such as Methane, Nitrous oxide, Chloro Fluoro Carbon (CFC) into the atmosphere by the human activities will raise the average global temperature of the atmosphere. This enhanced greenhouse effect is called global warming.

5.3.1. Effects of Global Warming

a. **Effect on Sea Level:** As a result of glacial melting and thermal expansion of the ocean, a 20 cm rise is expected in sea level by 2030.

b. **Effect on Agriculture and Forestry:** High CO_2 level in the atmosphere have long-term negative effects on crop production and forest growth. More grain belts would become less productive. As climatic pattern shifts, rain fall is reduced and soils are dried out resulting in major drought.

c. **Effect on Water Resources:** When the global rainfall pattern changes, the water management strategies of different regions have to adapt to these changes. Drought and floods will become more common, while rising temperature will increase domestic water demand.

d. **Effect on Terrestrial Ecosystems:** Many plant and animal species will have problems of adapting; this will influence the mix of species at different locations. Many will be at the risk of extinction, whereas more tolerant varieties will thrive.

e. **Effect on Human Health:** As the earth becomes warmer, the floods and droughts become more frequent. There would be increase in waterborne diseases, infectious diseases carried by mosquitoes and other disease vectors. The climate change might cause some ecosystems to exceed critical thresholds and result in irreversible decline.

5.3.2. *Preventive Measures*

The following measures will help to check Global Warming.

a. Emission can be cut by reducing the use of fossil fuels.

b. Implementation of energy conservation measures.

c. Utilization of renewable resources such as wind, solar and hydropower.

d. Planting more trees as they act as CO_2 sinks.

e. Shifting from coal to natural gas.

f. Adopting sustainable CO_2 agriculture.

g. Stabilizing population growth.

h. Efficient removal of CO_2 from smoke stacks.

Clean Development Mechanism (CDM)

CDM is an arrangement under the Kyoto protocol allowing industrialized countries with a greenhouse gas reduction commitment to invest in projects that reduce emissions in developing countries. It aims to develop sustainable development in all countries by reducing CO_2 and HFC (Hydro Fluoro Carbon) emissions.

5.4. Acid Rain

Acid rain means the presence of excessive acid in rain water. The thermal power plants, industries and vehicles release nitrous oxide and sulphur dioxide into atmosphere. When these gases react with water vapour in the atmosphere, they form acid and descend on to earth as "acid rain" through rain water. Normal rain water is always slightly acidic because of the fact that Carbon di oxide (CO_2) present in the atmosphere gets dissolved in it. Because of the presence of SO_2 and NO_2 gases as pollutants in the atmosphere, the pH of the rain water is further lowered. This type of acidic precipitation of rain is known as acid rain.

$$SO_x + H2O \rightarrow H2SO4 \quad NO_x + H2O \rightarrow HNO3$$

Due to the drifting of these gases in the atmosphere by the wind, their presence is felt as far as 2,000 kilometers. The air pollution of one nation could cause acid rain in another nation. Acid rain is also known as Acid fog, Acid snow and Acid precipitation.

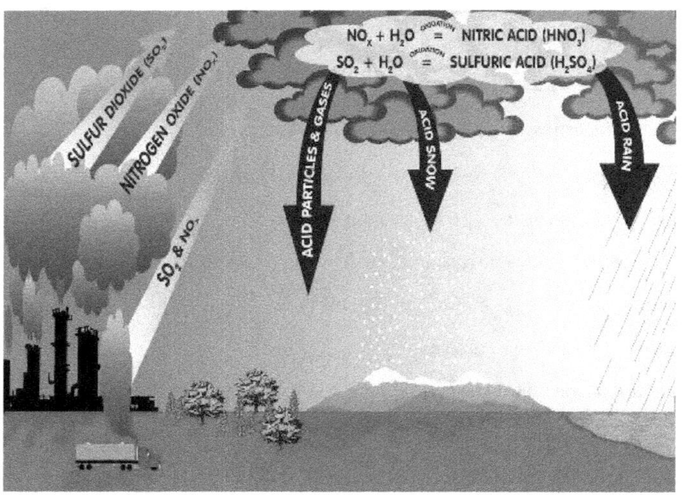

5.4.1. Effects of Acid Rain

Acid rain causes a number of harmful effects. Some of the adverse effects are as follows.

i.   Effect of Acid Rain on Human Beings

The toxic metals released due to acid rain can infiltrate into the drinking water and the animals and crops that human beings use as sources of food. This contaminated food can damage the nerves in children, or result in severe brain damage, or even death. Acidic rain has the ability of harming us via the atmosphere as well as the soil, where crops are grown and has been found to be very dangerous to human beings. It affects human nervous system, respiratory system and digestive system. It causes the premature death from heart and lung disorders.

ii. Effect of Acid Rain on Buildings

1. Acid rain corrodes houses, monuments, statues, bridges and fences made of marble, limestone, sandstone and deteriorate them in quality. The marble walls and pillars of the great man-made monument, Taj Mahal was found to be getting eroded by acid rain.

2. Acid rain and dry deposition of acidic particles contribute to the corrosion of metals, and the deterioration of paint and stone. These effects seriously reduce the value of buildings, bridges and cultural objects.

3. Dry deposition of acidic compounds can also dirty buildings and other structures, leading to increase of maintenance costs.

iii. Effect of Acid Rain on Terrestrial and Lake Ecosystem

1. The effect of acid precipitation on terrestrial vegetation reduces rate of photosynthesis and growth and increased sensitivity to drought and disease.

2. Acid rain severely retards the growth of crops such as beans, radish, potato, spinach and carrots etc.

3. Acid rain causes a number of complications in ponds, rivers, and lakes where it accumulates as acid snow. It causes a significant reduction in fish population.

4. The activity of the bacteria and other microscopic organisms are reduced in acidic water. So the dead materials and other accumulated substances lying on the bottom of lakes are not rapidly decomposed. Thus essential nutrients such as nitrogen and phosphorus stay locked up in dead wastages. Biomass production is reduced and fish population declines.

5.4.2. Control Measures for Acid Rain

1. Emissions of SO_2 and NO_2 from industries and power plants could be reduced by using pollution control equipments.

2. Coal with lower Sulphur content is desirable to use in thermal plants. Replacement of coal by natural gas could also reduce the problem.

3. Liming of lakes and soils should be done to correct the adverse effects of acid rain.

4. The real solution is to cut back on the use of fossil fuels by reducing our dependency on motor vehicles and unnecessary utility of motor articles.

5.5. Ozone Layer Depletion (Ozone Hole)

Ozone is a pale blue gas (O_3) with sweetish odor and unstable poisonous gas. Ozonelayer is the ozone rich zone between 10 – 50 km in the stratosphere but high concentration is around 20 – 25 km.

5.5.1. *Importance of Ozone Layer*

Ozone layer protects us from the damaging ultraviolet radiation of the sun as it prevents the entry of UV radiation to the earth. It prevents the disastrous radiation from different planets in the space and the earth. It reduces the temperature of the earth. It forms a pure atmosphere.

The studies have shown that certain parts of the ozone layer are becoming thinner and ozone 'holes' have developed. The consequence of thinning of the ozone layer is that more UV-B radiation reaches the earth's surface. UV-B radiation affects DNA molecules, causing damage to the outer surface of plants and animals. In humans it causes skin cancer and eye disease.

5.5.2. *Effects of Ozone Layer Depletion*

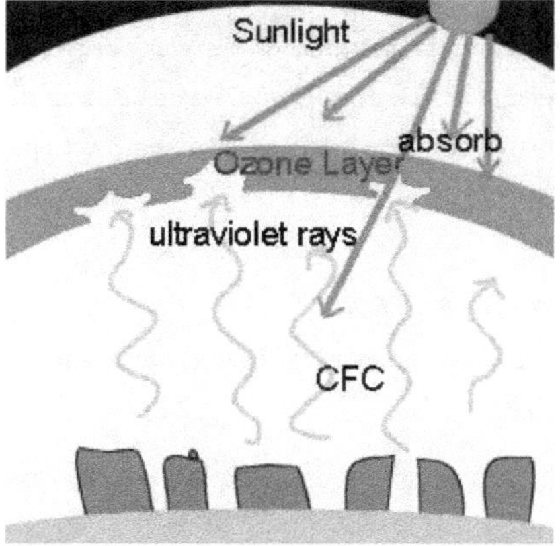

As the ozone layer gets thin, the harmful UV rays will reach the earth and cause various adverse effects.

a. *Effect on Human Health*

1. The UV rays damage genetic material in the skin cells which cause skin cancer.
2. Prolonged human exposure to UV-rays may lead to slow blindness called actinic keratitis.
3. Human exposure to UV-rays can suppress the immune responses in humans and animals. It also reduces human resistivity leading to a number of diseases such as cancer, allergies and other infectious diseases.

b. *Effect on Aquatic Systems*

1. UV rays directly affect the aquatic forms such as phytoplankton, fish, larval crabs, etc.

2. The phytoplankton consumes large amount of CO_2 Decrease in population of phytoplankton would lead to more amount of CO_2 in the atmosphere which contributes to global warming.

c. *Effect on Materials*

UV radiation causes the degradation of paints, plastics and other polymeric material that will result in economic losses.

d. *Effect on Climate*

The ozone depleting chemicals will contribute to the global warming.

5.5.3. *Control Measures*

1. Reducing the use of aircrafts, rockets, satellites and nuclear tests.

2. Use of Methyl Bromide, CFC, Carbon tetra chloride, nitric oxide can be minimized.

3. Strict licensing policies for opening of chemical industries.

World Meteorological Organization has declared 16th September as World Ozone Day to create awareness in the minds of people about dangerous ozone depletion. It is the duty of every individual to leave the earth as a paradise for future generations.

5.6. Environmental Legislation and Laws

Environmental management requires a strong framework in order to protect our valuable environment from the sources which are causing severe environmental pollution.

The major environmental problems are:

a. Air and water pollution by industries.

b. Deforestation.

c. Exploitation of Land resources.

d. Urbanization.

e. Waste management.

5.6.1. *Important Environment Protection Acts*

The main objective of the Protection Acts is to protect and preserve the eco system. Government of India and State governments have implemented a number of protection acts such as,

a. Water (Prevention and Control of Pollution) Act 1974.

b. Water (Prevention and Control of Pollution) Amendment Act, 1987.

c. Air (Prevention and Control of Pollution) Act, 1981 amended in 1987.

d. Wildlife (Protection) Act, 1972.

e. Forest (Conservation) Act, 1980.

f. Environment (Protection) Act, 1986.

a. Water (Prevention and Control of Pollution) Act, 1974

The main objectives of the water act are:

i) Prevention and control of water pollution.

ii) Maintaining or restoring the wholesomeness of water.

iii) Establishing central and state boards for the prevention and control of water pollution.

Important Features of Water Act

i) This Act aims at, to protect the water from all kinds of pollution and to preserve the quality of water in all aquifers.

ii) The Act further provides for the establishment of Central Board and State Boards for prevention of water pollution.

iii) The States are empowered to restrain any person from discharging pollutant or sewage or effluent into any water body without the consent of the Board.

iv) Any contravention of the guidelines or standards would attract penal action including prison sentence ranging from three months to six years.

The Amendment Act of 1987 requires permission to set up an industry which may discharge effluent.

State Pollution Control Board

The consent of the State Pollution Control Board is needed to,

i) Take steps to establish any industry or any treatment and disposal system or any extension or addition there to, which is likely to discharge effluent into a stream or well or river or on land.

ii) Use any new or altered outlet for the discharge of a sewage.

In the event of a violation of the conditions imposed, the State Board may serve on the offender a notice imposing any such conditions as it might establish, such outlet or discharge is a violation of the conditions.

The Act further empowers the State Board to order closure or stoppage of supply of electricity, water or any other services to the polluting unit. Non-compliance of the order may attract imprisonment for a term of one and half years to six years and fine which may extend to Rupees five thousand for every day, if the default continues.

b. Air (Prevention and Control of Pollution) Act, 1981

This Act deals with the prevention, control and abatement of air pollution. It envisages the establishment of Central and State Control Boards endowed with absolute power to monitor air quality and pollution control.

Objectives of Air Act

a. To prevent, control and abatement of air pollution.

b. To maintain the quality of air.

c. To establish a board for the prevention and control of air pollution.

Important Features of Air Act

a. The Central Board may lay down the standards for the quality of air.

b. The Central Board coordinates and settles disputes between state boards, in addition to providing technical assistance and guidance to State Boards.

c. The State Boards are empowered to lay down the standards for emissions of air pollutants from industrial units or automobiles or other sources.

d. The State Boards are to collect and disseminate information related to air pollution and also to function as inspectorates of air pollution.

e. The State Boards are to examine the manufacturing processes and the control of equipment to verify whether they meet the standards prescribed.

f. The State Boards can advise the State Government to declare certain heavily polluted areas as pollution control areas and can advice to avoid the burning of waste products which cause air pollution in such areas.

g. The directions of the Central Board are mandatory on State Boards.

h. The operation of an industrial unit is prohibited in heavily polluted areas without the consent of the Central Board.

i. Violation of law is punishable with imprisonment for a term which may extend to three months or fine up to Rupees ten thousand or both.

This Act is applicable to all polluting industries. The Air Act, like Water Act, confers wide powers on State Boards to order closure of any industrial unit or stoppage or regulation of supply of water, electricity or other services, if it is highly polluting.

c. Forest (Conservation (or) Preservative) Act, 1980

This act helps in the conservation of forests and related aspects. This Act is enacted in 1980 and covers all type of forests including served forests, protected forests and any forested land.

Objectives of Forest Act

a. To protect and conserve the forest.

b. To ensure judicious use of forest products.

Important Features of Forest Act

a. The reserved forests shall not be diverted or de-reserved without the prior permission of the central government.

b. The land that has been notified or registered or forest land may not be used for non-forest purposes.

c. Any illegal non-forest activity within a forest area can be immediately stopped under the act.

Important Features of Amendment Act of 1988

a. Forest departments are forbidden to assign any forest land 'by way of lease or otherwise to any private person' or non-government body for re-afforestation.

b. Clearance of any forest land of naturally grown trees for the purposes of re-afforestation is forbidden.

c. The diversion of forest land for non-forest uses is cognizable offence and anyone who violates the law is punishable.

d. Wildlife (Protection) Act, 1972

Wildlife (Protection) Act passed in 1972 and amended in 1983, 1986, 1991 and 2003. This act is aimed to protect and preserve wildlife. Wildlife refers to all animals and plants that are not domesticated. India has rich wildlife heritage. It has 350 species of mammals, 1200 species of birds and about 20,000 known species of insects. Some of them are listed as 'endangered species' in the Wildlife (Protection) Act.

Wildlife is an integral part of our ecology and plays an essential role in its functioning. The wildlife is declining due to human actions, the wildlife products – skins, furs, feathers, ivory etc., have decimated the populations of many species. Wildlife populations are regularly monitored and management strategies are formulated to protect them.

Objectives of the Wildlife Act

a.   To maintain essential ecological processes and life-supporting systems.

b.   To preserve biodiversity.

c.   To ensure the judicious use of species.

Important Features

a.   The act covers the rights and non-rights of forest dwellers.

b.   It provides restricted grazing in sanctuaries but prohibits in national parks.

c.   It also prohibits the collection of non-timber forest products.

d.   The rights of forest dwellers recognized by the Forest Policy of 1988 are taken away by the Amended Wildlife life Act of 1991.

e.   Environment (Protection) Act, 1986

This is a general legislation law in order to rectify the gaps and laps in the above Acts. This Act empowers the Central government to fix the standards for quality of air, water, soil and noise and to formulate procedures and safe guards for handling of hazardous substances.

Objectives of Environment Act

a.   To protect and improve the environment.

b.   To prevent hazards to all living creatures and properties.

c.   To maintain harmonious relationship between humans and their environment.

Important Features of Environment Act

1.   The Act further empowers the Government to lay down procedures and safe guards for the prevention of accidents which cause pollution and remedial measures if an accident occurs.

2.   The Government has the authority to close or prohibit or regulate any industry or its operation, if the violation of the provisions of the Act occurs.

3. The penal sections of the Act contain more stringent penalties. Any person who fails to comply or who contravenes any provision of the Act shall be punishable with an imprisonment for a term extending to five years or be punishable with a fine up to Rupees one lakh or both.

4. If the violation continues, an additional fine of Rupees five thousands per day may be imposed for the entire period of violation of rules.

5. The Act fixes the liability of the offence punishable under Act on the person who is directly in charge.

The Act empowers the officer of Central government to inspect the site or the plant or the machinery for preventing pollution; and to collect samples of air, water, soil or other material from any factory or its premises for testing. The Environment (Protection) Act is the most comprehensive legislation with powers for the central government to directly act, avoiding many regulatory authorities or agencies.

5.7. Environmental Ethics

5.7.1. Definition

"Environmental ethics refers to the issues, principles and guidelines relating to human interactions with their environment". Environmental ethics is a branch of ethics that studies the relation of human beings and the environment and how ethicsplay a role in this.

5.7.2. Functions of Environment

a. It is the life supporting medium for all organisms.

b. It provides food, air, water, and other important natural resources to the human beings.

c. It disintegrates all the waste materials discharged by the modern society.

d. It moderates the climatic conditions of the soil.

e. A healthy economy depends on a healthy environment.

5.7.3. Environmental Problems

a. Deforestation activities.

b. Demographic pressure (Population growth) and urbanization.

c. Pollution due to discharge of effluent and smoke discharge from theindustries.

d. Water scarcity.

e. Land degradation and degradation of soil fertility.

5.7.4. Ethical Guidelines

a. Worshiping and honoring the earth since it has blessed us with life and our survival.

b. Human being should not hold themselves above other living things and have no right to drive them to extinction.

c. Human beings must be grateful to the plants and animals which nourish us by giving food.

d. Man should not steal from future generations their right to live in a clean and safe planet by polluting it.

e. Sustainable utilization of the earth's precious treasure of resources.

Environmental ethics believe that humans are a part of society like any other living creatures, which includes plants and animals. These items are a very important part of the world and are considered to be a functional part of human life. Thus, it is essential that every human being should respect and honor the surroundings and use morals and ethics when dealing with these creatures. It is the responsibility of all to ensure that environmental ethics are being met.

5.8. Public Awareness

Our environment is presently degrading due to many activities like pollution, deforestation, overgrazing, rapid industrialization and urbanization. In order to conserve our environment, each and every one must be aware of our environmental problems and objectives of various environmental policies at national and local levels.

5.8.1. Objectives of Public Awareness

a. To create awareness among the rural and urban people about ecological imbalances, local environment, technological development and various development plans.

b. To organize meetings, group discussion on development, tree plantation programmes, exhibitions.

c. To focus on current environment problems and situations.

d. To train planners, decision-makers, politicians and administrators.

e. To eliminate poverty by providing employment that overcome the basic environmental issues.

f. To learn to lead a simple and eco-friendly life.

5.8.2. *Methods to Create Environmental Awareness*

Environmental awareness must be created through formal and informal education to all sections of the society. The various methods that are useful for increasing environmental awareness are:

1. **Environmental Awareness in Schools and Colleges:** Environmental education must be imparted to the students in schools and colleges.
2. **Mass-media:** Media like Radio, TV and cable network can educate the people on environmental issues through cartoons, documentaries, plantation campaign, street plays.
3. **Cinema:** Films about environmental education should be prepared and screened in the theatre compulsorily. These films may be released with tax free to attract the public.
4. **Newspapers:** All the newspapers as well as magazines must publish the environment related problems.
5. **Audio-Visual Media:** To disseminate the concept of environment, special audio – visual and slide shows should be arranged in all public places.
6. **Voluntary Organizations:** The services of the voluntary bodies like NCC, NSS, should be effectively utilized for spreading the environmental awareness.
7. **Traditional Techniques:** The traditional techniques like folk plays, dramas, may be utilized to spread environmental messages to the public. These techniques attract the rural people very much.
8. **Arranging Competitions:** Story writing, essay writing and painting competitions on environmental issues should be organized for students, as well as for the public. Attractive prizes should be awarded for the best effort.
9. **Leaders Appeal:** Political leaders, cine actors and popular social reformers can make an appeal to the public about the urgency of environmental protection.
10. **Non-Government Organizations (NGOs):** Voluntary organizations can help the government to solve the local environmental issues. Also they can be effective in organizing public movements for protection of environment through creation of awareness.

CHAPTER VI

CERTIFICATION OF AUDIT

Introduction of 'Eco Audit', 'Green Audit', 'Hygiene Audit' and 'Energy Audit' to Institutions, Organizations and Industries to improve the eco-friendly campus and to reduce the environmental pollution. The audit reports and recommendations may be useful for providing a hygienic environment to the stakeholders.

The audit process can be classified into four types. They are:

a. Green campus audit

b. Environment audit

c. Energy audit

d. Hygiene audit

6.1. Audit Process

Based on the checklists prepared as per the ISO 9001: 2015 along with the Government Law and Environmental Legislation, the audits are being carried out in the form of observations, report preparation and corrective actions are being taken up. Institutions, Organizations and Industries who are willing to undergo audits.

6.2. Green Campus Audit

Green auditing is a type of assessing environmental performance with respect to maintaining the green campus by planting trees. To ensure that the list of plants and animals / birds available in the campus, biodiversity conservation, water irrigation system, recycling of water, rain harvesting system, site preservation, soil erosion control and landscape management will be evaluated. This includes all emissions to air; land and water; legal constraints, the effects on the neighboring community; landscape and ecology; the public perception of the operating company in the local area. Green audit does not stop all compliance with legislation. Nor is it a 'green washing' public relations exercise, rather it is a total strategic approach to the organizations activities.

6.3. Environment Audit

First concept of Environment Audit was approved in the Earth Summit, Rio – 1992. In Institutes practice like renewable energy varies from 60% to 30 %, sewage treatment plants only 20%. Green audit is a tool of management system used methodologically for protection and conservation of Environment and sustenance of Environment. Green youth Leadership is to conserve and comprehend the relationship with the plants, animals. Eco-audit or Environmental audit is a process of extracting information about an Institution and Organization that provides a realistic assessment of how the campus surroundings affect the environment and make sure that their activities should not harm the environment. This audit can minimize the environmental pollution in the campus which in turn reduces the global warming effect as a whole.

6.4.   Energy Audit

Energy Audit ensures that the campus of Institutions and Organizations is involving energy savings and consumption towards the road-map of the State development economy by assessing the electric current usage. As per the Energy Conservation Act, 2001, Energy Audit is defined as "the verification, monitoring and analysis of use of energy" including submission of technical report containing recommendations for improving energy efficiency with cost benefit analysis and an action plan to reduce energy consumption.

An energy audit is a survey in which the study of energy flows for the purpose of conservation is examined at an organization. It refers to a technique or system that seeks to reduce the amount of energy used in the organization without impacting output. The audit includes suggestions of alternative means and methods for achieving energy savings to a greater extend. Conventionally, electrical energy is generated by means of fossil fuels, hydraulic and wind. The need for an energy audit is to identify the savings potential and cost reducing methods, understand the ways to which fuel is used, where; the waste occurs and find the scope for improvement.

6.5.   Hygiene Audit

A Hygiene audit will provide an insight into an organization operates in a sustainable manner in terms of hygiene environment to the stake Holders as per International standard for Occupational Health and Safety Management Systems (ISOHSMS). If an organization has a hygiene auditing process implemented already, then it should apply environment Hygiene audit is a process of extracting information about hygienic environmental context into a clean environment. Hygiene audit provides a realistic assessment of how the Institutions and Organizations cause adverse effect to human beings health and surroundings. To ensure that the hygienic management system, maintenance of environmental and personal hygiene, cleanliness ensured at the site of disposal of hazardous materials, personal safety, microbial load (bacteria, fungi and viruses) in the surroundings are being implemented effectively. An effective hygiene practice should be followed among the stake holders which in turn useful to control a wide variety of disease outbreaks in the environment.

A healthy population is the essential component of a country's wealth political, economic and environmental sustainability.

Chapter VII

Activities Planned for Awareness, Attitude and Green Practices among Students

Environmental attitude of young people appears to be crucial as they ultimately play a direct role in providing knowledge based solutions to incoming environmental problems. Since the present problems result largely from ignorance, there is the need for attitudinal and behavioral change, commitment in people which is essential in effective participation in waste reduction, reuse and recycling. Below are some suggestions for easy-to-do activities with a sustainability theme for the under graduate students. Each activity is designed to develop the skills such as observing, analysing, creating, problem-solving, and decision making.

Objectives of the Activities

- To change the attitude of the students this will in turn influence behavioral change.
- To train the students to develop skills to solve environmental problems.
- To make the students Green Consumers & Green Entrepreneurs.
- To make the students creative and pave way to express their green thoughts.

Activities for Interactions with Flora & Fauna

Activity 1: Animal Diversity of Campus (Campus Tour)

Students have visit the campus and take geo-tagged photographs of animals like like birds, butterflies, spiders, amphibians, insects, mammals and and make a document.

Objectives:

- To understand the adaptations / mimicry / different larval stages of organisms.
- To understand the relationship among organisms.

Activity 2: Seed Ball Preparation

Seed balls are seeds that have been wrapped in soil matter and then dried. The soil matter is often a mixture of clay and compost. The wrapping mixture keeps the seed safe until it can properly germinate. With seed balls, we can propagate plants and trees from seeds, without the need for opening the soil with cultivation tools like a plough. When sufficient rain permeates the clay casing and given the conditions are favorable the seed will germinate. Seed balls are one of the easiest and sustainable ways of cultivating plants.

Activity 3: Gardening (Green Creativity) & Organic Farming

Gardening is a part of horticulture and landscaping which is mostly done for increasing aesthetic beauty or making the environment healthier. A College garden will provide students a gain in insight on how to grow a garden. Students can involve in Gardening at College campus (Planting & Watering / Maintaining) and maintaining a specific type of garden / implement innovative gardening techniques.

- Using pet bottle as planters / Using Tender Coconut shell as planters.

- Vertical gardening / indoor plants in pet bottles (Exhibition / Sale within campus).

Activity 4: Creating / Setting a Wildlife-friendly Garden at Home

An array of colorful nectar-rich flowers will attract bees, wasps, butterflies and other insects. Providing bird boxes, constructing artificial nests, Giving feed and water to birds can be a great way of introducing good artificial shelters and food into nature.

Objectives

- To Create a wildlife-friendly garden and to make the gardens be home for wildlife.

- To introduce, or let nature create, a diverse range of homes for nature in outdoor spaces.

- To Grow fruit trees (natural habitat & food for birds and animals).

Green Entrepreneurial Activities

Activity 5: Paper Pen Making / Paper Bead Making (Upcycling / Reducing Plastic Usage)

To make paper pens, the refills are covered with used paper. Use of single use plastic pens reduces plastic pilling up landfills.

Outcomes

- Students become Green Entrepreneurs & they earn while they learn.
- Got opportunity to exhibit their creativity and they indirectly motivate the society to become green consumers.

Feminine Hygiene Products Making

Sanitary napkin disposal is a daunting challenge, one that causes health and environmental hazards. Annually, 12 billion sanitary napkins are used in India, of which 98% ends up in water bodies and landfills. It takes 800 years for a single sanitary napkin to decompose. The training will be given to the students to stitch *Reusable Cloth pads* on their own and to make *eco-friendly biocompostable hygienic products*. Cotton Sanitary Napkins can be made using cotton, non-woven fabric by pressing it using Iron Box or by stitching it.

Objectives: To use a reusable product this enables the production of less trash and emission of less carbon dioxide at the time of manufacturing it.

Outcome: After learning the skill, the students will be able to become an entrepreneur and they will be able to earn. By using these products we are preventing the sanitary napkins piling up in landfills. The re-usable cloth napkins protect the user from coming in contact with a number of harmful chemicals found in disposable napkins.

Activities to Realize & Practice 'Reducing, Reusing & Recycling (3R)'

Activity 6: Field Visit & Report on "Landfill Areas and Highly Polluted Areas Around Me – My Observations and Solutions"

The students can visit a few landfill areas and highly polluted areas and identify the contributors of landfill area and also visualize the real problems and threats of dumping wastes.

Objectives

- To learn about the types of wastes.
- To find the ways to cut down on the waste they produce and they realize the importance of segregating the waste and disposing.
- To make the students realize the importance of "Reducing, Reusing, and Recycling (3R)".
- To make the students understand the importance recycling the biodegradable wastes to reduce the amount of solid waste reaching landfills.

Activity 7: Composting in Transparent Pet Bottles (Recycling Biocompostable Wastes)

Wastes from kitchen or garden can be used for making compost. Pet bottles (preferable) or any tubs can be used as a container. Every week observation has to be recorded.

Objectives: To give students hands on experience on composting technology, this enables them to get practical knowledge in recycling biodegradable solid waste into valuable compost.

UPCYCLING / REUSING ACTIVITIES

Upcycling, also known as **creative reuse**, is the process of transforming by-products, waste materials, useless, or unwanted products into new materials or products perceived to be of greater quality, such as artistic value or environmental.

It is essential to understand why any disposable things can be harmful to the environment. One should consider the carbon emissions generated during the manufacturing process of a disposable things and its active lifespan, apart from waste generation, treatment and disposal techniques.

Activity 8: Upcycling Unused Things - Wealth from Waste

Creative recycling activities at home can help save the planet. When the plastic bottles & other wastes are not discarded, the damage done to ocean life is also reduced.

Objectives

- To showcase the innovative thoughts of the students.

- To turn their recycling ideas into money-making projects (Businesses around upcycling items).

- To make the students to think innovatively and find solutions to use any type of wastes and upcycle into better products.

Eg. Conversion of T-shirts into bags, Tooth brush Bracelets, using tires for making seats.

Activity 9: Ecobricks Making

Ecobricking is a simple way to take personal responsibility for our plastic by keeping it out of industry and out of the biosphere. An ecobrick is a PET bottle packed solid with CLEAN and DRY used plastic. Ecobricks are made manually to a set density to sequester plastic and create reusable building blocks.

Objectives

- To keep plastic from degrading into toxins and microplastics by ecobricking.

- To deepen our awareness of these issues by hands-on process of ecobricking.

Activity for Sharing Green Thoughts and Innovative Ideas

Activity 10: Recorded Video / Poster Making

A talk can be given (Recorded Video) by the students on environmental issues / how to address it / green practices from literature & History / Importance of Commemorating national and international days / or any other related topics. Creative common pictures can be used in PPTs.

Objectives

- To Exhibit skills and Green ideas of the students and to educate and motivate others.
- To post in YouTube channel and creating awareness through media.